SCI论文写作和发表：
You Can Do It
（第三版）

[美]张俊东　　
杨亲正　国防｜编著

化学工业出版社
·北京·

内容简介

《SCI 论文写作和发表：You Can Do It》（第三版）先介绍了 SCI 论文各部分如标题、摘要、简介、结果、讨论等的具体写法和注意事项，然后介绍了投稿和发表过程中的期刊选择问题以及投稿信和答复信的具体写法，这两部分均附有若干具体示例，可作为模板使用。后半部分对英文科技写作中的常用语法、词汇进行了具体剖析和举例，以切实提高读者的写作水平。

本书可供高校理工农林医类专业研究生和广大的科研工作者参考使用。

图书在版编目（CIP）数据

SCI 论文写作和发表：You Can Do It：英、汉 /（美）张俊东，杨亲正，国防编著. —3 版. —北京：化学工业出版社，2023.6（2025.2重印）

ISBN 978-7-122-43136-3

Ⅰ．①S… Ⅱ．①张… ②杨… ③国… Ⅲ．①科学技术-论文-写作-研究-英、汉 Ⅳ．①G301

中国国家版本馆 CIP 数据核字（2023）第 049270 号

责任编辑：宋林青　汪 靓	装帧设计：关　飞
责任校对：宋 玮	

出版发行：化学工业出版社（北京市东城区青年湖南街 13 号　邮政编码 100011）
印　　装：河北延风印务有限公司
710mm×1000mm　1/16　印张 13　字数 247 千字　2025 年 2 月北京第 3 版第 4 次印刷

购书咨询：010-64518888　　　　　　　　售后服务：010-64518899
网　　址：http://www.cip.com.cn
凡购买本书，如有缺损质量问题，本社销售中心负责调换。

定　　价：45.00 元　　　　　　　　　　　　　　版权所有　违者必究

我们的烦恼（第一版序）

俊东是高我一级的学长，也是我的好朋友。记得当年在北医药学系读书时与俊东熟悉是在1984年暑假去山东的夏令营中，经过十几天的"同居"，我俩性格相近，又是山东同乡，自然就成了无话不谈的哥们。以后俊东和我先后在药学系不同的课题组读硕士，后来俊东考了个很高的TOEFL成绩去美国攻读博士学位去了，之后我们的联系也就少了！我在硕士二年级参加考试转博，博士毕业后就留校任教。从讲师、副教授到教授，2001年被聘为博士生导师，按时晋升，没受到丝毫挫折。慢慢有了自己的课题组，招收了若干研究生，也在国内糖化学界有了一点知名度。和众多"海龟"相比，我应该算是"混"得好的"土鳖"了。"土鳖"虽然生命力比较强，也服水土，但也有许多烦恼，缺钱、缺空间、招生名额太少等，其中一个不大不小的烦恼是如何把自己的工作写成SCI文章并发表在相应档次的期刊上。

中国人学习英文应该是最认真、花费时间也最久的，但"实用性"却不理想。我们许多人就是这种"全民英语工程"的产品，语法很熟，词汇量也不小，对学生写出来的论文改了又改，但投出去在内容得到肯定后，经常会得到一句"The manuscript has language and grammar issues and would benefit from an editing in scientific English"的评价，这种情况一直伴随着我们，影响着论文发表的速度、档次，有时甚至会被拒稿。2008年俊东回国探亲，我们又团聚了，了解到俊东在业余时间干了两件事。一是成立了一个专业的编译团队，帮助国内的学者修改、润色论文，并且成绩斐然。这对我简直是一个天大的喜事，解决了我们课题组的大问题。几年来在俊东及其团队的帮助下，我们组的SCI论文发表速度明显加快，档次稳步提升，数量也上来了，现在每年能在20篇以上。更重要的是，通过俊东的帮助，我和学生们在SCI论文的写作上也有了不小的收获，几名博士生通过修改、发表论文，写作水平得到提升，赴美留学后论文的撰写基本没有问题。

英文论文写作有它的规律和技巧，要达到专业的水平需要长期的训练和时间，并不是语言好就一定能写出好文章，中文论文的写作也是同样的道理。我了解到许多"海龟"甚至我们眼中很高的"大牛"们其论文也要请外籍人士或专业编辑来修改。在国际上，请专业编辑人才修改英文写作是学术界已普遍接受的一个通常做法，这一点国内也逐渐在接受。

俊东做的另一件事是"著书立说"，他结合自己多年 SCI 论文写作、修改和润色的经验，写了这本"SCI 论文写作和发表：You Can Do It"。俊东是个学者，他的专业横跨了化学、药学和生物学，他做过研究，发过文章。从中国到美国的经历，使他能了解中国学者英文写作的瓶颈在哪里！除了帮助大家修改论文外，他还打算从年轻人的培养做起。现在国内的青年学生都有很好的英文水平，如果能有一本实用、针对性强的手册指导，假以时日，经过几轮实战的训练，定能写出与研究水平相当的 SCI 论文发表！

俊东做了一件十分有意义的事情！

李中军

2015 年 12 月 18 日于北大医学部

序 言（第二版序）

张俊东、杨亲正先生和国防女士合编的《SCI论文写作和发表：You Can Do It》是一本面对国际专业期刊的科技论文英语写作实用指南，可以作为科技领域研究人员的必备参考书。它在多年来不断完善，并证明了它是一本很有用的工具书。

中国近现代的科技发展经历了漫长的、艰苦的积累过程，现在已经拥有坚实的基础和广大的从业者队伍，也不断涌现着本土创造。这时候我们就具备了与国际科技界和学术界平等对话的基本条件。这种对话，既是要参与国际范围内的科技发展和学科建设并作出自己的贡献，也是要继续向国外同行学习以不断充实我们自己的力量。必须从这样的双重角度认识为国际期刊写作科技论文的意义，并付出耐心的努力。至于功利目的，诚然也是正当的；但我们写作论文绝不能局限于狭隘的功利主义眼光（更不能抱着急功近利的草率态度），那将无法保证自身的可持续发展。毋庸讳言，不少人尚未在这个问题上有明确而坚定的认识。

要参与国际科技界的学术对话，就要了解对话的基本要求。这种要求表现为一系列的学术规范，它们是国外学者多年沿用并且已经相对固定下来的，我们一般只能遵守。此外，我们必须使用外语来写作，这又是一个横加给我们的负担。对于中国人来说，英语和日语大概是最容易学习的外语了；即便如此，中国人在学习和使用中还是有很大的困难。其他外语就更不用说了。若无专业的外语训练，要使自己的外语达到较高水平是很不容易的。在这方面，外语专业的学生（尤其是研究生）所受过的外语论文写作训练，是一种很有价值的财富。但是，他们一般并不学习科技的内容，也不从事科技论文的写作。我提及这一点，是想建议一些外语专业人士也来关心（甚至投身）科技论文的写作，至少主动协助科技工作者进行外语论文的写作。也想建议科技工作者随时找身边可以找到的外语工作者帮忙；更想建议科技工作者主动积极地学习外语并在运用中超过那些并无"长技"的专业外语工作者。现在新一代人外语水平普遍比老一代人高得多，上面这个目标并非不能实现。总之，大家应该怀着一个宗旨：要超越自己，要为中国、为世界多做贡献。

此书是特地为科技专业人员定制的"常规武器"。编著者张俊东博士用自己的亲身经验展示了他如何逐渐养成较高的英语科技论文写作能力的。此书在相当程度上不啻是他的现身说法。固然，他有在美国学习和工作的良好条件，但关键之事还在于他在论文写作（以及批改）中的反复历练。因此我认为，我们的科技工作者在任何时候都要把研习和揣摩外文科技论著（尤其是目标刊物上的优秀论文）列为头等重要的学习任务。与此同时，也可以（和应当）利用这本写作指南来指导和帮助自己更快更好地进行学习、模仿和创造。

正如张俊东博士所言，最重要的是我们在科技上必须真有自己的创新，这是一个前提。如果成果是过硬的、有创新的内容，就可以环绕这个中心点来组织自己的文章，并尽量以清晰的思路表述出来。当然，理论的依据、论证的步骤、实验的数据支持等方面，文中需要协调编排；具体到论文的章节、段落甚至小标题，也要有清晰和稳妥的措置，亦即必须注意语篇（包括句、段）之（文意的或语义内容的）连贯和（语法结构形式的）衔接。后面这种具体的技巧性工夫是与外语基本功紧密相关的，需要不断仔细揣摩体会、逐步积累经验；而在整个过程中一定要自觉地保持对"中式外语"的警惕性，才能渐入佳境，真正达到能运用规范、地道的外语进行自由撰述的目的。

如何有计划、有步骤地进行这种写作的自我训练，书中已经有很详细的指点，我就不多说了。愿广大读者能借助此书的阶梯稳步上进，不断实践、领悟和掌握，把自己锻炼成为科技论文外语写作的能手，在参与国际科技发展中、在提升中国科技实力中多作贡献。

北京师范大学外国语言文学学院

语言学教授　周流溪　谨识

2015年9月10日

前　言

我在美国学习和做科研已有35年之久了。来美后常收到国内朋友请求，让我帮助润稿。我深切体会到英文写作仍是国内学者的一个难题，于是开始研究中国学者，包括自己，英文写作的瓶颈在哪里。我自己英文写作的提高经历了一个漫长的过程。出国前我TOEFL考了当时北京的最高分数，这个分数让我争取到了美国常春藤大学的奖学金。但来美后英语仍是"听不懂，说不出，写不清"。我的MIT博士后导师是美国微生物学杂志主编，也写教科书。我的每篇论文他都仔细修改，使我受益匪浅。他的言传身教对我的写作水平提高起到了很大作用。

英文科技写作有它自己的规律，在一定的英文基础上，是可以达到专业水平的。我主要从三方面研究英文写作：一是阅读有关论文写作的专著；二是从写作的角度阅读大量不同杂志的优秀论文；三是通过修改国内学者的稿件，找出问题所在。这本书就是集中了过去25年我对英文科技论文写作研究的总结。

我在2006年组建Sciwriting编译团队（www.sciwriting.cn），已经协助国内学者修改15000多篇SCI稿件，结合各位编辑的意见，我发现国内学者SCI写作的问题主要表现在：

1. 词汇缺乏　导致用词不准或错误，不能准确表达。这可能是最大的问题。

2. 语法问题及流畅问题　科技写作的语法与基础英语的语法一样，但有些是科技英语中常用的语法问题，如时态、名词罗列等，我们对一些英语特有的语法掌握不好。论文写作思路要清晰，语句表达要流畅，但有时我们对语句的连接，思维的逻辑性表达不足。

3. 中式英语　中文的句式和语法与英语有很多不同之处，直接逐句翻译往往会写成中式英文。由于词汇不足和对英语写法的不了解，写出的句子会不符合英语习惯，这可能是最难克服的弱点。

而一些没发表的稿件除了存在上述问题，更多的是对SCI论文理解不深。SCI论文中创新好比是根本，好的思路是枝干，语言好比是枝叶。很多作者没有跟踪文献，创新性和意义无法提高，只在语言上下功夫，会舍本逐末。学英文写作是个漫长过程，不可能短时间内人人都成为专业人才。要增强国际交流，首先要把精力放在选好创新性题目，设计并完成实验上。英文写作要学好还需要花很大功夫，也需要一个能了解英语母语表达方式的环境。对于语言，请专业编辑人才修改英文写作是学术界接受的一个普遍做法。通过仔细研究专业人才所修改的稿件，学习SCI论文写作技巧，不断积累，才能逐步得到提升。

《SCI论文写作和发表：You Can Do It》第一版和第二版出版后，很受读者欢迎。同时热心的读者也提出了宝贵的改进建议。结合论文润色服务中了解到的作者需求，第三版增加了如下内容：

1. 添加了新的一章：投稿，稿件重修和答复信的写法。着重讲解如何重修稿件，特别是如何写好答复信。对如何答复一些常见到的审稿人的问题，详细举例说明。

2. 扩展了第三章中的"中式英文"部分，对一些常见的中式英文予以剖析。希望读者在自己的写作中尽量避免中式英文。

3. 对书中的大段英文，有的加了解释，有的进行了翻译，以便于读者阅读。

再次感谢读者的鼓励和支持。

张俊东

bosi@sciwriting.cn

2023年3月于Boston

Introduction

I have been studying and working in U.S. for more than 35 years. Frequently, my friends in China have asked me to edit their manuscripts. It was these requests that prompted me to study scientific writing in more depth and to look into what was necessary for Chinese scientists to prepare their own manuscripts in professional English. It took me a long time to improve my own English writing skills, even though I had a very high TOEFL score when I was applying US universities. My initial training was from my postdoctoral advisor at MIT, the editor-in-chief of a renowned biology journal as well as the author of several textbooks. By studying his revisions of my own manuscripts, my scientific writing skills improved greatly.

English writing, like many other skills, is mastered by both learning and practice. I further improved my skills by reading many scientific writing books and journal articles, while always paying special attention to the writing. Editing manuscripts of fellow Chinese scientists and identifying the common mistakes provided me additional insight. I published a book 《Guide for Scientific Writing》 by Shandong University Press Co. in 2004 to summarize the key areas that I believe are essential for Chinese scientists to improve their scientific writing skills. To meet the requests for manuscript editing, I built a team of editors in the U.S. to offer editing service in 2006 (**www.sciwriting.cn**). We have edited more than 15000 manuscripts so far. Enriched by the editing experience and authors' feedback, I have revised the book and published the 2nd edition in 2016. In this 3rd edition, a new chapter on "Manuscript Submission, Revision, and Response Letter Writing" is added. The section on how to avoid writing "Chinese English" is expanded, and more Chinese translations are provided for easier reading.

There are several obstacles Chinese scientists have to overcome. The first and most difficult one is the proper selection and usage of words. The poor writing of non-English speaking scientists can often be improved by developing knowledge of the subtle differences in the accurate use of words. Simple words are overused, and in many cases, the wrong words are used instead. The second obstacle is grammar and sentence flow. Scientific writing should meet the same standards for grammar as general writing while accommodating for the additional rules of scientific writing. The sentence flow, which can be improved by the correct usage of connecting words, is critical for readers to understand the authors. The third obstacle is Chinese- English: English written as if literally translated from Chinese, including inappropriate expressions and improper grammar. While the author might not realize, a native speaker can notice it right away.

Writing a publishable manuscript is a difficult task: one which requires hard work and persistence. A manuscript written in proper English is the prerequisite for its acceptance for publication; however, the most critical criterion is still the contents of the manuscript, especially the novelty of your research finding. To write clearly the purpose of your study, what questions you want to answer or hypothesis to demonstrate, and the differences from the previous reports are the core of the paper. Without a proper understanding of the requirements for a publishable manuscript in SCI journals and the up-to-date knowledge of the literature in your research field, even the best writer still cannot publish a scientific paper. It takes a long time, the proper language environment, and tremendous efforts to master the skills of English writing; it seems to me it is a better way for scientists to focus on their science and ask professional editors to help with their English writing. It is a well-accepted practice in English-speaking countries to have professional editors to edit a manuscript or grant proposal. It is this need for English editors that is driving me to sharpen my own English writing skills, train other interested individuals, and provide professional service for the needed scientists.

I sincerely appreciate the encouragement and support received from readers and fellow scientists.

Jundong Zhang
bosi@sciwriting.cn
March 2023, Boston

目 录

第 1 章 稿件的撰写 /001

1.1 如何写好一篇稿件 ···002
1.2 Title Page Including Authors, Affiliation and Contact Information ····004
    1.2.1 Title 设计时常见的内容性问题 ·······················006
    1.2.2 Title 设计时的可读性问题 ···························008
    1.2.3 Title 中常见的语法问题 ·····························009
1.3 Abstract and Keywords ································012
    1.3.1 Abstract 常见的三种格式 ···························012
    1.3.2 Abstract 常见问题 ·································013
    1.3.3 不同格式 Abstract 示例 ····························013
1.4 Introduction ··019
1.5 Materials and Methods or Experimental Section ········030
1.6 Results ··033
1.7 Discussion ···039
1.8 Conclusion ··043
1.9 Acknowledgments ·····································043
1.10 References ···044
1.11 修改稿件 ··045

第 2 章 稿件投稿和发表 /049

2.1 期刊选择和稿件投稿 ···································050
    2.1.1 稿件评价标准 ······································051
    2.1.2 Cover letter 的写法 ································053

- 2.2 稿件重修 054
- 2.3 答复信的写法 055
- 2.4 一些常见问题的答复范例 058
  - 2.4.1 对要求补充实验或数据的答复 058
  - 2.4.2 对要求改进章节中表述的答复 063
  - 2.4.3 对要求改善实验或数据的答复 067
  - 2.4.4 对要求改正不准确语言的答复 070
  - 2.4.5 查重率太高问题 072

第3章 英文写作 /075

- 3.1 论文写作中常出现的语法问题 076
  - 3.1.1 主语和谓语的单数和复数要一致 076
  - 3.1.2 修饰语同主语名词关系上要一致 077
  - 3.1.3 主语和主语的行动（谓语）在逻辑上要一致 079
  - 3.1.4 代名词和其代理的先行词要一致 080
  - 3.1.5 位置的强调作用 081
  - 3.1.6 修饰词和被修饰词要邻近 082
  - 3.1.7 主语和谓语在句子中的位置要靠近 083
  - 3.1.8 名词作形容词 084
  - 3.1.9 句子的时态 085
  - 3.1.10 主动句和被动句 086
  - 3.1.11 标点符号的使用 086
  - 3.1.12 数字的写法 087
  - 3.1.13 冠词的使用 088
  - 3.1.14 同位词的使用 089
  - 3.1.15 多余的用词 089
  - 3.1.16 隔离 090
  - 3.1.17 Units 090
- 3.2 句子的连接和信息的传承 091
  - 3.2.1 信息的传承 092

3.2.2　句子的连接 ·· 093
　　3.2.3　平行句的组织方法 ·· 096
　　3.2.4　写简单句子 ·· 097
3.3　中式英文 ·· 099

第4章　常用词汇 /107

4.1　经常使用但容易出现问题的字 ·· 110
4.2　用以描写研究课题的意义和重要性 ·· 119
4.3　用以描述某个领域的现状或作者的研究计划 ·· 122
4.4　用以描述需要解决的问题 ·· 127
4.5　有关过去和现在时间的词汇 ·· 131
4.6　用于举例的词汇 ·· 133
4.7　用以描述实验部分的词汇 ·· 136
4.8　用于描写先后顺序的词汇 ·· 142
4.9　描写实验结果的词汇 ·· 144
4.10　用于比较语句的词汇 ·· 155
4.11　用于讨论和描述实验意义的词汇 ·· 158
4.12　常用的修饰词 ·· 168

附：稿件样本 /175

附录 /185

References /191

第1章 稿件的撰写

1.1 如何写好一篇稿件

每人都有自己喜欢的写作方式，并没有一个统一的最好办法。但写好一篇稿件所面临的问题都是一样的。一篇好的 SCI 论文需要回答下面四个问题：

为什么（why）要做这个研究（前言部分）；

如何（how）来做实验（实验部分）；

实验结果是什么（what）（结果部分）；

科研结果的意义、贡献和亮点是什么或与众不同之处在哪里（what）（讨论部分）。

刚开始论文写作时会感觉无从下手，建议按以下顺序着手。

① 阅读和积累文献，特别是阅读实验室过去发表的文章，主要是查证自己的科研是否解决了一个文献上没有解决的问题，是否增进了对课题的进一步认识，自己科研的特殊意义在哪里。对自己的设想，也需要查文献看是否合理。同时了解论文格式，描述方法，领域的前沿和面临的问题等。

② 写结果部分。先从自己熟悉的部分开始写作，比如先写结果部分。这部分内容自己熟悉，也是一篇论文的基础内容。结果能用图表（Figure and Table）表述的，最好做成图表。图表的质量很重要，也需要时间去整理。在结果整理过程中会常常产生一些新的想法，也许会发现一些数据还不是很全面，可再做些补充实验，从而及时将遗漏的数据补充完善。

③ 写实验方法部分。实验方法要同实验结果匹配，要针对实验结果部分写。不能有遗漏，也不能有多余的实验。一些实验参数、试剂要及时记录下来，以免信息丢失。

④ 写前言部分。这时对自己科研结果的意义、创新有数了，比较容易写前言部分的 why, how, and what。写前言部分需要有一些文献的积累，要准确引用文献，最好引用原始论文。

⑤ 写讨论部分。讨论部分与引言部分应该是衔接的，都涉及文献的背景知识和本论文的贡献。讨论应该在已有文献的知识下展开，阐述论文与以往报道的不同、对研究课题的贡献和作者的看法等。

⑥ 写结论和摘要。不是每个期刊都要求有结论部分。论文主体写好之后，写

结论和摘要就容易了。

⑦ 写论文标题 Title。

⑧ 对整个论文反复修改。充实稿件内容，完成草稿。修改稿件时应先从整体上开始，看内容是否完整，思路和段落是否合理，段落之间是否要调整顺序。找导师、朋友帮助修改，提意见，主要是内容方面的。

⑨ 修改英语。仔细阅读每个句子，主要看是否通顺，句子与句子之间是否有合理的连接。而后再检查用字是否准确，特别注意全文用词要一致，字词的输入是否有错误，标点符号的使用是否规范，文献是否准确等。修改时每遍最好只集中注意某个内容，这样效果好，也不容易遗漏。必要时，可寻求专业人士修改英语写作。

⑩ 仔细阅读期刊的 Guide for Authors。投稿前，按期刊的要求整理文献、摘要和其他主体部分。

与一般文学写作不同，SCI 论文稿件有一定的格式，尽管不同 SCI 期刊一般又有自己的特殊要求，但基本格式是一致的。稿件的行与行之间要留有手写修改文字的空间，最好用 double space。文字一般用 12 号字，Figure 和 Table 都是附在稿件的后面，与文字分开，而不是安插在文字中间。稿件一般要有以下几个部分，按先后顺序为：

Title Page Including Authors, Affiliation and Contact Information (one page)

Abstract and Keywords (one page)

Introduction

Materials and Methods

Results

Discussion

Conclusion

Acknowledgments (and Author Contribution, etc.)

References

Table (one table per page)

Figure Captions

Figure (one figure per page)

为满足不同期刊的特殊要求，读者应先阅读投稿说明（Guide for Authors）并参阅期刊上已发表的论文。

写好科学论文的基本条件是首先作者要思路清晰，对自己的科研领域、实验结果和它的意义，有清楚的了解；否则，再好的语言表达能力也无用武之地。其次，运用自己的英语能力，通过合理的语法，正确的字词拼写，组织好完整的句子和清晰的段落。论文要一句一句地写，写作是一个需要下很多功夫的过程。写作中的用词、组句、文献和数据的阐述，都要经过多次修改才能达到满意的程度。所以要预留足够时间来整理论文，一般要 3 个月，越早动笔越好。下面介绍每个部分的写法和内容要求。

1.2 Title Page Including Authors，Affiliation and Contact Information

Title 可以说是文章中最重要的一句话，它应简要地表达出文章的内容。一个好的标题，即具有简练醒目，引人入胜的特点，又概括了整个文章的内容，可以吸引读者阅读。考虑到以上因素，设定标题，首先要考虑内容性。SCI 论文标题需要围绕着研究对象、研究方法和研究结果三个部分或至少两部分来设计。例如，"Preparation and antibacterial activity of lysozyme and layered double hydroxide nanocomposites"，包含了研究对象"lysozyme and layered double hydroxide nanocomposites"和研究结果"preparation and antibacterial activity"。其次，要考虑可读性。即使内容上概括了整个文章，但如果表达错误，或太啰嗦，也会影响审稿人对稿件作出正确评价。

标题要简洁清晰，内容性强，用字尽量少。可以写成概括阐述的形式，如：

Identification of the Substrate Binding Sites of Protein K

Design and Synthesis of Potent Caspase Inhibitors

也可以写成具体描述的形式，如：

Residue Arg-123 and Leu-67 Are Involved in the Substrate Binding of Protein K

Benzo[3, 4]diazepines as Potent Caspase Inhibitors

当用概括阐述的形式时，注意标题应与文章的内容相符合，不能用太宽泛的描述，这样具体描述的形式才能更明确和信息性强。读者应参考自己的科研领域的文献，注意学习其他作者是如何写 Title 的。

除非是第一个词，标题中的冠词（a，an，the），介词（at，in，on，of），连接词（and，but，if）的第一个字母不大写，如：

The Dependence of Crystal Growth on the Solvents

写作者和工作单位时，中文名字用汉语拼音，要写全名，不要简写，一般应名在前，姓在后。双名的拼音一般要写在一起，如 Likang Zhang；若把两个字的拼音分开，其中应加连字符，如 Li-Kang Zhang。中文名字的英文写法比较混乱，除了大陆和港台的拼法不同外，大陆也经常有不同的人用不同的写法。无论采用哪种写法，最重要的是保持自己名字的英文写法一致。

作者一般按贡献大小排列，第一作者是主要实验操作者，最后一个通常是指导教授。但也有教授把自己的名字列为第一作者的。如果作者中有多人做出了同等贡献，尤其是第一作者，可以加注解符号备注出来，这样就成为并列第一作者。若作者来自不同科研单位，一般在名字右上角加注解，如 1、2、3 等，但是要慎重标记，一般标记 1 的为第一单位。联系人的名字右上角一般加*号，也称为通讯作者。有的期刊在申请后可以并列通讯作者，标记两个加*号作者。Title 页的下半页写联系地址、电话和 E-mail 地址等。

常见的 Title 页：

Rat Plasma Stability Study of Insulin by LC-MS

Laiwen Liu[1], Ming Wang[2], and Qikan Zhao[1*]

1. Department of Chemistry, Beijing University, Beijing 100021, China

2. Department of Biological Sciences, Brown University, Providence, RI 02102, USA

[*]Corresponding author

Contact Information

Qikan Zhao

Department of Chemistry

Beijing University

Beijing 100021

China

Email：

标题设计时一般容易犯三个方面的错误：一是写得太简单，没有有效地反映稿件的内容；二是写得太庸长，太啰嗦；三是语法错误。其中语法错误最常见。

1.2.1　Title 设计时常见的内容性问题

例 1.

Delivery of DNA Vaccine Based Chitosan Nanoparticles for Oral Immunization against Reddish Body Iridovirus in Turbots *(Scophthalmus maximus)*

标题中研究对象是 DNA vaccine，研究结果是 delivery。经过通读发现文章中不仅研究了 delivery，还研究了制备、分析等内容，这样如标题中仅用 Delivery 就不能很好地概括稿件的研究内容。同时，标题中使用了 oral immunization 词组，oral 是服用的一种方式，在英文中无法和 immunization 连起来使用。针对以上问题，将研究对象改为 oral DNA vaccine，研究结果改为 Development。整个标题宜改为：

Development of Oral DNA Vaccine Based on Chitosan Nanoparticles for the Immunization against Reddish Body Iridovirus in Turbots *(Scophthalmus maximus)*

例 2.

Transcriptome Sequencing, Annotation and Expression Analysis of Brain Tissue at Different Gonad Maturation Period and Reproductive Regulation-related Genes of Starry Flounder (*Platichthys stellatus*)

这个标题写得太多、太具体。研究对象是 brain tissue 和 starry flounder (星斑川鲽)，范围较大，实际为 hypothalamus (丘脑下部) and pituitary (垂体) gland，研究结果是 transcriptome (转录组)、annotation (注解) and expression analysis，其实应为 transcriptome profiling。而且 different gonad (性腺) maturation period and reproductive (生殖的) regulation-related genes 是两类不同概念，不能用 and 并列连接。整个标题宜改为：

Transcriptome Profiling of the Hypothalamus and Pituitary Gland at Maturation and Regression Phases in Starry Flounder (*Platichthys stellatus*)

例 3.

Global Sensitivity and Protection Measures Analysis of the Parameter Influenced Pipeline Risk Induced by Metro Deep Pit Excavation

标题中研究对象是 Parameter，研究方法是 Global Sensitivity Analysis，标题中用了两个动词，读者不易理解。并且还表达不清楚. 通读稿件后，加上研究的目的 for Protective Measure Design，整个标题宜改为：

Global Sensitivity Analysis of Pipeline Safety Risk Factors Adjacent to Metro

Deep Pit Excavation for Protective Measure Design

例 4.

The Study of Newborn Screening and Genetic Mutation for Phenylketonuria in Weishi

这个标题的缺点是太宽泛，特征性不强，信息量不足。研究对象是 infants with phenylketonuria（苯丙酮酸尿症），研究结果是 newborn screening（筛选）and genetic mutation（变异）analysis。但对于哪些人群筛选没有描述清楚，同时研究地点交代也不明确，没明确写出在中国。根据文章表达内容，整个标题宜改为：

Newborn Screening of 0.7 Million Infants in Fifteen Years and Genetic Mutation Analysis of the Infants with Phenylketonuria in Weishi City of China

例 5.

In Vitro and In Vivo Evaluation of Group-10 Metal 6-Amino-oxoisoaporphine Complexes. Selective Stabilization of G-Quadruplex DNA, Inhibition of Telomerase Activity and Disruption of Mitochondrial Functions

这个标题中，研究对象是 Metal and 6-Amino-Oxoisoaporphine Complexes，研究结果 In Vitro and In Vivo, Stabilization of G-Quadruplex DNA, Inhibition of Telomerase, Disruption of Mitochondrial。标题中研究结果概括性差，可以统一理解成生物活性研究。同时，标题应该是一个句子，很少写成两个句子。整个标题宜改为：

Preparation and Evaluation of the Biological Activities of Group-10 Metal and 6-Amino-oxoisoaporphine Complexes

例 6.

Characteristics of Eformation and Ailure

标题太简单，没有提到研究对象，而其单词也写错了。根据文章表达内容，整个标题宜改为：

Characteristics of Deformation（变形）and Failure（破坏）and Mining（采矿）Pressure of Deep Coal Seam Floor（煤层底板）Affected by Fully Mechanized Mining

例 7.

Dual-phase Ni-Al-Fe Ternary Intermetallics Synthesis by Combustion Method under High Pressure and Their Hardness Property

本标题研究对象为 Ni-Al-Fe ternary（三元的）intermetallics（金属间化合物），研究方法 combustion（燃烧），研究结果为 synthesis 和 hardness。但标题中没有

把这三者关系表述清楚，Dual-phase Ni-Al-Fe ternary intermetallics synthesis 应改成 Synthesis of dual-phase Ni-Al-Fe ternary intermetallics，对于性质研究，应该是 study of their hardness。整个标题宜改为：

Synthesis of Dual-phase Ni-Al-Fe Ternary Intermetallics by Combustion Method under High Pressure and Study of Their Hardness

1.2.2　Title 设计时的可读性问题

例 8.

The Protective Effect of Fisetin on Ischemia/Reperfusion Myocardial Cells and Its Mechanism in Rats

这个标题的问题在于句子结构以及语法让人读起来很繁琐，主题不明确。经修改后的标题简洁且明确，整个标题宜改为：

Protective Effects of Fisetin（非瑟酮）in a Rat Model of Myocardial Ischemia and Reperfusion（心肌缺血和再灌注）

例 9.

Inhibiting Effects of *Bacillus subtilis* on *Microcystis aeruginosa*

Effects 是个多余的字，应该去掉。即

Inhibition of *Microcystis aeruginosa*（微囊藻毒素）by *Bacillus subtilis* （枯草杆菌）

例 10.

Mechanisms of Gastric Carcinoma: *Helicobacter Pylori* Virulence Factors

这个标题只列出了两组名词，之间的相互关系是什么要靠读者自己体会。标题应该更明确地点出两组名词间的关系：

Helicobacter Pylori（幽门螺杆菌）Virulence （毒性）Factors in Development of Gastric Carcinoma（胃肿瘤）

例 11.

Identification of Characteristic and Prognostic Value of Chromosome 1p Abnormality by Multi-gene Fluorescence in situ Hybridization in Multiple Myeloma

这个标题写得含糊，不知道具体意思是什么。Characteristic value 作者的意思是 characterize。value 也是一个国内作者经常使用的词，但它的含义很广，不知道具体是想说什么，应该避免使用，类似的还有 situation、performance、scientific base、 phenomenon 等。整个标题宜调整为：

Using Multi-gene Fluorescence（多基因荧光）in situ Hybridization（原位杂化）to Identify and Characterize Chromosome（染色体）1p Abnormality as a Prognostic（预后）Factor in Multiple Myeloma（多发性骨髓瘤）

例 12.

Mesenchymal Stem Cell Transplantation Combined with Interleukin-1 Receptor Antagonist Benefit Liver Regeneration after Hepatectomy by Balancing the Signaling Network of Inflammation and Apoptosis

这个标题的缺点是太繁琐。作用的机理 by balancing the signaling network of inflammation and apoptosis 可以省略。mesenchymal stem cell transplantation 是中文直译，翻译成英文应该是 transplantation of mesenchymal stem cell。标题调整后为：

Transplantation of Mesenchymal（骨髓间质）Stem Cells in Combination with Interleukin-1（白细胞介素）Receptor Antagonist（拮抗药）Benefits Liver Regeneration Post Hepatectomy（肝切除）

例 13.

Manipulative Synthesis of Multipod Frameworks of Cu_2O Microcrystals and Cu_7S_4 Hollow Microcages

没有 manipulative synthesis 这一用法，改为 synthesis and characterization，标题宜调整成：

Synthesis and Characterization of Multipod Frameworks of Cu_2O Microcrystals and Cu_7S_4 Hollow Microcages

1.2.3　Title 中常见的语法问题

例 14.

The mir-590-3p is a Novel microRNA in Myocarditis by Targeting Nuclear Factor kappaB in vivo

这个标题语法结构不合理，targeting nuclear factor kappaB 是修饰 microRNA 的，被 in myocarditis 隔开了，in vivo（体内），是多余的。标题的优点是简洁，切中稿件的要义。调整后为：

Mir-590-3p is a Novel microRNA Targeting Nuclear Factor kappaB in Myocarditis (心肌炎)

例 15.

The Advances of Midkine With Peripheral Invasion In Pancreatic Cancer

这是一篇综述的标题。一般综述的标题要简洁，且能概括比较大的范围。而一般实验方面的论文特征性要强，这个标题的缺点是语法不合理，peripheral invasion 是 midkine 作用机理的一部分，应该直接写出 biological mechanisms，而不写具体机理。With 和 In 应该首字母小写，标题宜调整为：

Biological Mechanisms of Midkine (肝素结合细胞因子) in Pancreatic (胰腺) Cancer

例 16.

iTRAQ Proteomic Approach Reveals Salinity Adaptation Protein in the Gill of the Tropical Marbled eel (*Anguilla marmorata*)

reveals salinity adaptation protein 不知道具体是什么意思，还是属于语法或句子结构问题，标题宜调整为：

iTRAQ proteomic（蛋白质组）Analysis of Salinity Adaptation Proteins (盐度适应蛋白) in Gill (鳃) of Tropical Marbled eel (热带花鳗鲡)(*Anguilla marmorata*)

例 17.

Efficient Visible-light Photo-catalytic Hydrogen Evolution and Enhanced Photostability for Core@Shell $Cu_2O@$ $g-C_3N_4$ nanospheres

标题就是把多个名词罗列起来，很繁琐。用 visible-light 来修饰 photo-catalytic hydrogen evolution 也不合理，应改为 visible-light induced hydrogen evolution。介词 for 用得也不对，应该用 of，改为 enhanced photostability of 或 photostable $Cu_2O@$ $g-C_3N_4$ nanospheres 都可以。标题调整后为：

Efficient Visible-light Induced Hydrogen Evolution Catalyzed by Photostable $Cu_2O@$ $g-C_3N_4$ Nanospheres

例 18.

Large Temperature Coefficient HDPE/BST Polymer-Ceramic Composites for Wireless Temperature Sensation Application

名词修饰语的用法不对。large temperature coefficient 是三个词，用来修饰 HDPE/BST polymer-ceramic composites，到底谁修饰谁，很混乱。整个标题应调整为：

HDPE/BST Polymer-Ceramic Composites（聚合物陶瓷复合体）with Large

Temperature Coefficient for Wireless Temperature Sensation（无线温度感应）。

例 19.

Stable Ag_3PO_4@g-C_3N_4 Core/Shell Hybrid Structure with Enhanced Visible Photocatalytic Degradation

因为只报道一个 composite, 应加冠词 a。degradation 应该用 activity 代替。整个标题宜改为:

A Stable Ag_3PO_4@g-C_3N_4 Hybrid Core/Shell Composite with Enhanced Visible Light Photocatalytic Activity

例 20.

Lipid Profile Analysis in One of the Edible Jellyfish *Rhopilema Esculentum* by UPLC-ESI-Q-TOF-MS

这个标题的问题也是语法。lipid profile analysis 应该是 analysis of the lipid profile，介词 in 用得也不对，标题修改后为:

Analysis of the Lipid Profile of Edible Jellyfish（海蜇）*Rhopilema Esculentum* by UPLC-ESI-Q-TOF-MS

例 21.

5-Fluorouracil (5-FU) Pharmacokinetic and Pharmacodynamic Analyses in East-Asian Patients with Nasopharyngeal Carcinoma (NPC)

名词作修饰语的用法不对。中文是可以说"5-FU 的药代动力学"，英文应该写成"药代动力学 of 5-FU"，标题修改后为:

Pharmacokinetic and Pharmacodynamic（药代动力学和药效学）Analyses of 5-Fluorouracil (5-FU) in East-Asian Patients with Nasopharyngeal Carcinoma（鼻咽癌）(NPC)

例 22.

Cervical Sagittal Alignment in Chinese Adolescent Idiopathic Scoliosis patients

标题中，名词修饰语的用法不当，中文可以写成 adolescent idiopathic scoliosis patients，英文翻译应该是 patients with adolescent idiopathic scoliosis。标题中其他词都开头大写，patients 也应该开头大写。标题中第一个字母大写或小写都可以，但要一致。标题修改后为:

Cervical Sagittal Alignment（颈椎矢状排列）in Chinese Adolescent（青少年的）

Patients with Idiopathic Scoliosis（特发性脊柱侧弯）

例 23.

Preparation of the Porous Superabsorbent Polymer from Chitosan and Application for Hemostatic Dressing

标题中的介词用得不对。from chitosan 应该是 porous superabsorbent polymer of chitosan，或 chitosan porous superabsorbent polymer。for hemostatic dressing 应该是 as a hemostatic dressing。标题修改后为：

Preparation of Chitosan Porous Superabsorbent Polymer（多孔的超强吸附的壳聚糖聚合物）and its Application as a Hemostatic Dressing（止血敷料）。

设计标题时，除了要注意以上问题，在阅读文献时，多留意已发表文章的标题，对自己设计标题会有很大帮助。对于语法问题，只要作者把内容把握好，可考虑先拟出中文标题，然后请教专业语言处理人员，使标题变成能准确表达自己思想的英文 Title。

1.3 Abstract and Keywords

Abstract 要单独占一页。虽然 Abstract 在论文的前面，但一般都最后写。稿件主体部分写成后，Abstract 就容易写了。

Abstract 就是论文的简本，浓缩了论文最重要的部分，能让审稿人或读者在不读全文的情况下了解作者的科研结果，来判断是否阅读全文及是否接收或引用该篇论文。因此，一篇论文的主要内容，如课题背景、要解决的问题、实验设计、关键结果和结论都应包括在 Abstract 中。期刊对摘要字数一般都有严格要求，100 到 200 字不等，目的是言简意赅。一般都是稿件主体部分写成后，才来写 Abstract。Abstract 中会有结论性一小句话，但要避免和 Conclusion 中的句子重复。

Keywords 一般要写出 5 个左右与文章最相关的词用以供检索时使用，可以单独一页，也可以写在 Abstract 的后面。

1.3.1 Abstract 常见的三种格式

摘要的格式一般可以分为三类，一段式、结构式和图表式。

一段式，是用一个段落把实验目的，方法，结论和总结论述清楚。这是最常见的格式。一般生物、工程、数学、物理、社会科学等都采取这个方式。

结构式，就是期刊要求按一定结构撰写，主要是一些医学期刊。一般要求要有 Purpose, Design, Methods, Results, Conclusion。也有期刊要求写 Significance。

图表式，主要是用于化学期刊，要求把反应方程，化学结构或材料结构图像等列入摘要。一个图胜似上百字的语言，有了图后，文字部分一般都很简短，50～100 字即可。

1.3.2 Abstract 常见问题

英文摘要写作中易出现的问题一是英文写作错误，二是对摘要的写作内容掌握不好。期刊编辑和审稿人一般是先阅读摘要，当发现摘要语句不通，内容表述不清时，一般就会考虑退稿了。而一个语句通顺、表达清晰的摘要会给读者一个好印象。经常出现一些作者把论文主体写得很漂亮，但认为摘要简单，自己整理后就投稿了。期刊发现头两句就有语法错误，要求润色语言。这样不仅耽误了不少时间，还给审稿人和主编留下差的第一印象，很是可惜。另一个常见的问题是不够简练，也就是字数太多。编辑部要求缩减和作者要求帮助缩减的情况很多。在写作中，不妨先写一个较长的摘要，然后细心研究每一句，决定是否可以去掉多余的表达内容。

1.3.3 不同格式 Abstract 示例

虽然不同的期刊对摘要的格式有不同的要求，但在内容方面，要求是一样的。对初学英文论文写作的作者，可以简单地把摘要限制在 10 句。用第一句来写科研背景和目的，对一些目的明显的科研，如化学类研究，这一句可考虑省略。第二句概括性地写本论文做了什么。接下来用 3～4 句话写实验方法，再用 3～4 句写实验结果，最后一句写总结或意义。

下面用一些实例来介绍摘要写作中一些常见的问题。

例 1.

Abstract: It is a promising treatment strategy to use a nanoparticle-based drug delivery system for cancer patients, which can simultaneously deliver multiple drugs or genes in combination with therapy to induce synergistic effects and suppress drug resistance to the tumor (第一句写科研背景). In this study, cationic nanostructured lipid carriers (cNLC) for co-loading anionic small-interfering RNAs (siRNA) and chemotherapeutic docetaxel (DTX) were prepared from different cationic lipids based on particle distribution and loading efficiency (第二句写本论文做了什么). A novel peptide SP94 was bound to the surface of cNLC (SP94-cNLC) (这一句写实验方法,

有些简短,只写了材料制备,没写生物实验方法). The cNLC showed good efficiency in loading siRNA and DTX. The SP94-cNLC revealed a better cytotoxicity compared with cNLC and Taxotere®, indicating that SP94 could successfully enhance the internalization capacity of nanoparticles to the liver cancer cells. This new type of cNLC is a potential vehicle when using in co-delivery of chemotherapeutics and siRNAs (这句结论应放在最后,不应与实验结果混在一起). The curcumin (CUR)/DTX co-delivery NLC could load both CUR and DTX in high efficiency and showed a sensibilization to DTX chemotherapy. The sensibilization was more obvious when it was used in the aggressive and resistant cancer cells (有四句写实验结果,比较合适). This CUR/DTX co-delivery system had good potential in treating cancer cells when chemotherapy drug showed little effect alone (最后一句写出结论和意义).

这段摘要的科研背景和具体做了什么工作有点冗长,科研方法部分不足,实验结果和结论比较合适。

例2.

Abstract: PEGylation, the covalent attachment of polyethylene glycol (PEG) polymers, is often used for improving the pharmacological properties of therapeutic proteins. PEGylation is usually achieved by tethering PEG to the reactive group of constitutive amino acids. However, it often results in a complex mixture of conjugates with different number and distribution of bound PEGs (先写科研背景,有些长,中间一句可以去掉). Here, we develop a genetic code expansion method as a general approach for the selective PEGylation of INF-α2b at any chosen site(s) (再写本论文做了什么). Our approach involves two key steps: incorporation of an azide-bearing amino acid through genetic code expansion as the attaching site into INF-α2b; orthogonal conjugation of a variety of PEGs with INF-α2b, site-specifically and stoichiometricaly, via a strain-promoted alkyne-azide cycloaddition (这一句是写实验方法). By this approach, homogenous PEGylation at only the chosen site(s) is consistently obtained, thus avoiding non-selective conjugations (这一句写实验结果). Furthermore, our approach provides a facile platform for optimization of the PEGylation of INF-α2b via different sizes of PEG at different sites (这一句应放在最后,是总结性的一句). The higher biological activities associated with H34 and E107 positional isomers and the improved PK profile by a single 20-kDa PEG modification may provide insights for the development of next generation of therapeutic INF-α2b (这一句像实验结果的描述,最好放在实验结果部分).

这段摘要科研背景有些长,科研方法部分比较合适,实验结果部分有些短,最后的结论一句更像实验结果的描述,可与前一句调换一下。

例3.

Abstract: *Caenorhabditis elegans* (*C. elegans*) was used as an animal model to study the effect of (–)-5-hydroxy-equol, a microbial metabolite of isoflavone genistein, on the lifespan, fecundity and resistance against thermal and oxidative stress (第一句写本论文做了什么，并带有实验方法). The results showed that (–)-5-hydroxy-equol not only significantly increased the lifespan of *C. elegans* but also significantly enhanced the resistance against thermal and oxidative stress at the concentrations of 0.1 mmol/L and 0.2 mmol/L. However, the fecundity of *C. elegans* was not obviously influenced after being exposed to the same concentrations of (–)-5-hydroxy-equol. Further study using comparative transcriptome analyses and of the effect on *daf-16* (mu86) mutant and *daf-2* (e1370) mutant indicated that (–)-5-hydroxy-equol prolonged the lifespan of *C. elegans* through DAF-2/DAF-16 Insulin/IGF-1 signaling pathway (用三句话描述实验结果). This is the first report that (–)-5-hydroxy-equol was able to increase the lifespan and improve the thermal and oxidative stress tolerance of *C. elegans* (最后一句写出结论和意义).

这段摘要写得简单明了，非常合适。

例4.

Abstract: Sphingolipids, a new class of lipid mediators, are involved in a variety of important physiological and pathological processes. Sphingomyelin synthase (SMS) is an enzyme to convert the ceramide and phosphatidylcholine into sphingomyelin and diacylglycerol, which plays a key role in sphingolipid biosynthesis. Two SMS isoforms, SMS1 and SMS2, have been identified with different subcellular localizations and expression level in tissues. Previous studies have shown that SMS may serve as a potential therapeutic target for the treatment of various diseases, such as cardiovascular and metabolic diseases. Thus, there is an urgent need for a rapid and sensitive method for SMS activity analysis (用了5句话写科研背景，占了几乎一半的字数，太长了，应保留一句，或最多两句). In our study, we developed a novel method for SMS activity with five steps (i.e. homogenate preparation, incubation with substrate, lipid extraction, thin-layer chromatography (TLC) separation, and NBD intensity quantification), and applied this method in cells and mice.(这一句写本论文做了什么，并带有实验方法). In Huh7 cells, the interassay coefficient of variation of the SMS activity assay was $(3.6\pm0.07)\%$. In wild type (WT) mice, we observed accumulation of NBD-sphingomyolin in blood in a time dependent fashion. In SMS 2 KO mice, NBD-sphingomyolin in plasma collected 5 (0%, $P<0.01$), 30 (16%, $P<0.01$), and 60 min (21%, $P<0.01$) after injection of fluorescence liposome solution with D609 was significantly decreased compared with WT mice. However, in SMS1 KO mice, NBD-sphingomyelin in plasma collected 5- and 30 min is similar to that in WT mice.(用 4 句话描述实验结果). Our results suggest that this method could be used for SMS activity measurement in vitro and in vivo. (本句写出结论). More studies in other samples are needed to confirm our conclusion (本句应该去掉，或放讨论部分).

这段摘要科研背景太长，科研方法和实验结果部分比较合适，结论一句也很好。最后讨论的一句应该删掉或放在讨论部分。

例 5.

$$\text{unactivated alkyl halides} \xrightarrow[\text{dioxane, 100℃}]{\text{10 mol \%Pd(PPh}_3)_4,\ 2\text{equiv K}_3\text{PO}_4} \text{82\% isolated yield diverse carbocycles and heterocycles}$$

A catalytic C–H alkylation using unactivated alkyl halides and a variety of arenes and heteroarenes is described (第一句写本论文做了什么). This ring-forming process is successful with a variety of unactivated primary and secondary alkyl halides, including those with β-hydrogens. In contrast to standard polar or radical cyclizations of aromatic systems, electronic activation of the substrate is not required (中间描述实验结果). The mild, catalytic reaction conditions are highly functional group tolerant and facilitate access to a diverse range of synthetically and medicinally important carbocyclic and heterocyclic systems (最后一句写出结论和意义).

这段摘要是图表格式 (Graphic Abstract)，由化学反应式和简短的文字描述组成，是一个很典型的化学合成论文摘要。

例 6.

The chirality transfer of axially chiral binaphthyl derivatives bearing liquid crystal (LC) moieties at the n,n' positions ($n = 3, 4, 6$) of the binaphthyl rings to nematic (N) and smectic (S) LCs is investigated (第一句写本论文做了什么). Chiral nematic LCs (N*-LCs) are prepared by adding a small amount of the chiral binaphthyl derivative into host N-LCs composed of cyanobiphenyl mesogen cores. The binaphthyl derivative with phenylcyclohexyl (PCH) type LC moieties at the 4,4′ positions of the binaphthyl ring [**D-4,4′**] exhibits a low helical twisting power (HTP) of 11 μm^{-1}. In contrast, those with LC moieties at the 3,3′ and 6,6′ positions of the binaphthyl rings [**D-3,3′** and **D-6,6′**] exhibit high HTPs of 153 μm^{-1} and 154 μm^{-1}, respectively (中间

描述实验结果)。It is concluded that **D-6,6′** has a large helical twisting power and is the most favourable atropisomeric chiral inducer for chirality transfer to both N-LCs and S-LCs (最后一句写出结论)。

有的化学期刊不采用图表格式，而用一段式。

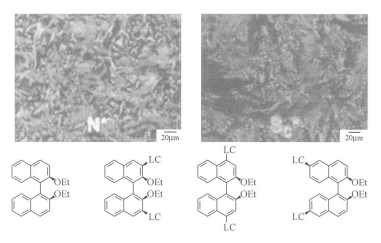

例7.

Abstract: To gain insights into the direct electron transfer (DET) mechanism of multi-walled carbon nanotubes (MWCNTs), we investigated the conformational changes that occur in proteins when they interact with MWCNTs (第一句写本论文做了什么). We used glucose oxidase (GOD) as an example. Using cyclic voltammetry measurements, the GOD that was immobilized on the MWCNT-modified carbon paper electrode exhibited apparent direct electrochemistry compared to that on the bare electrode without MWCNTs. The structural transformation of GOD upon adsorption on the MWCNTs was characterized spectrally (这几句写实验方法). GOD was not denatured, and only small shifts of the wavenumber of the b-sheet structure were observed. There was a consistent tendency for the amount of a-helix to decrease and the b-sheet to increase. The a-helix content dropped from 21.2% to 19.6% as measured using Fourier transform infrared spectroscopy and from 27.1% to 25.9% as measured using circular dichroism. The reduction in the amount of a-helix led to a less shielded GOD active site and weakened the resistance of the electron transfer (再写实验结果). These MWCNT-induced conformational changes could account for the DET between GOD and the MWCNT-modified electrode surface (最后写总结).

一些医学期刊采用结构式摘要，要写的几个方面都给固定好了。这样写起来比较容易，也方便控制字数。下面为实例。

例 8.

Abstract

Objective: To study the differences in palate growth morphology between children with anterior crossbite and children with normal occlusion at deciduous stage（用一句描述论文研究的目的）.

Materials and Methods: 3-D measurement analysis method and mathematical polynomial fitting were used to obtain the palate growth morphology of 126 Han children with normal occlusion and 118 Han Children with anterior crossbite in Xi'an and analyze their differences based on their ages. Triangular interpolation method was used to obtain and compare the average surfaces of palate growth images of the 3~5-year-old children in these two groups（简要描述重要的材料和实验方法，一般不超过三句）.

Results: (1) The developments of palate growth of children with anterior crossbite were limited in both sagittal and coronal planes with their age increasing compared with that of children with normal crossbite; (2) The symmetry of palate growth of children with crossbite was worse than that of children with normal occlusion; (3) Although the surface difference analysis showed that the morphological reconstruction of children with normal occlusion and children with crossbite at deciduous stage was not effective, the changes in palate surface growth of children with crossbite appeared to be more retarded（科研成果和结论，要详细，定量，有重点，结果部分应该是最长的。这个例子写得有些短，只是文字描述，没有定量数据）.

Conclusion: Deciduous anterior crossbite limits the development of palate morphogenesis at three dimensions（用简洁的一句描述论文的结论）.

例 9.

Abstract

Purpose: In 1984, the German Breast Cancer Study Group started a multicenter randomized trial to compare six versus three cycles of cyclophosphamide, methotrexate, and fluorouracil (CMF) starting perioperatively and to investigate the additional effect of tamoxifen as adjuvant treatment in node-positive breast cancer patients treated with mastectomy（科研目的有些长，应该考虑缩减一半）.

Patients and Methods: From 1984 to 1989, 473 patients were randomized from 41 institutions. After a median follow-up of approximately 10 years for overall survival (OS) and 9 years for event-free survival (EFS), the treatment groups were compared with respect to OS and EFS. Results based on a median follow-up of 56 months have been published earlier（患者和方法写得清晰，字数也合适）.

Results: Estimated cumulative locoregional incidence rate after 10 years was 19.9%; the corresponding rate of distant recurrences was 41.3%. Concerning duration of chemotherapy, we did not find any difference between six and three cycles of CMF (EFS: relative risk [RR] in multivariate analysis = 0.95; 95% confidence interval [CI], 0.74 to 1.21, OS: RR = 0.90; 95% CI, 0.69 to 1.18). Treatment with tamoxifen resulted in an improvement in outcome (EFS: RR = 0.81; 95% CI, 0.61 to 1.07, OS: RR = 0.74; 95% CI, 0.55 to 1.0), although it proved not significant. Number of positive lymph nodes and progesterone receptor were the dominant prognostic factors（结果部分详细，有定量描述，字数也合适）.

Conclusion: In this study, we observed some tendency in favor of hormonal treatment, which is in agreement with the literature. Concerning duration of chemotherapy, the results of this study provide further evidence that a reduction to three cycles of CMF is possible without increasing the risk of recurrence or death. For a definitive conclusion; however, further investigations are required（结论部分有些长，应考虑缩减）.

1.4 Introduction

Introduction 部分要对论文研究的领域做一简要介绍，要涉及研究的意义，现在的问题在哪里，你是如何去解决问题的。一般先从大面上说起，然后再转到你研究的课题，再阐明现在的问题是什么，及你是如何去研究这个问题的。这些一般都是比较简要的描述，详细的探讨通常放在 Discussion 部分进行。 在 Introduction 部分要回答以下几个问题，并把这几个问题作为粗略提纲列出来。通过文献检索，根据自己的知识积累，逐步丰富：

① 某个领域现在是什么情况；

② 研究这个课题的意义；

③ 存在的问题是什么；

④ 本文是如何来进一步深入研究的。

从字数上来看，主要部分是文献综述。这需要较宽广的文献知识，并且每个阐述都要有文献来源。比如你写了十个句子来描述课题已有的知识，至少要有十个以上的文献来源。文献要与自己的科研有关，在写作时，手头上最好有 10～20 篇与课题密切相关的文献。阅读这些有关文献，看其他作者是如何组织科研背景介绍的。

描述研究本课题的必要性时，一句或最多两句话就可以。根据实验结果写前言，有目的地写，需要什么样的新结果。结尾可以用三句话：我们做了什么，发现了什么和科研的意义。

Introduction 开头的几句话要使用标题中的关键词，以便使文章直接进入主题，把读者引入到你的研究课题中去，不要写与主题无关的字句。比如文章的标题是"A New Strategy to Improve the Production Yield of Insulin"，开头的第一句话中就应有 Insulin。接下来就简要介绍课题的背景知识，也就是描述文献研究。然后提出现在的问题或缺陷在哪里，作者是如何进一步去研究的。

Introduction 的写法上最常出现的有两种格式。一种是先描述某个领域的进展情况，再转到存在的问题，然后阐述作者是如何去研究和寻找答案的。现举一个简单的例子：

Heat-shock proteins (HSP) are induced when a cell undergoes various types of environmental stresses like heat, cold and oxygen deprivation. They are involved in a variety of biological functions, such as protein folding, translocation, and immune presentation.

……（中间添加具体文献研究结果，然后转入现在的问题是什么）

However, the exact mechanism of how they exert their biological functions remains unclear. In this study, we carried out kinetic experiments to investigate the conformational changes in the structure of HSP's substrates.

另一种是直接从描述研究的课题的意义下手，然后阐述作者是如何去研究的，比如：

Early diagnosis is critical in cancer treatment and cure. In the case of breast cancer, the survival rate was increased dramatically when therapeutic drugs were administered in the early stages of the cancer.

……（中间添加具体文献研究结果）

To address the need for early diagnosis of breast cancer patients, we have developed two sensitive and specific immunoassays to qualify protein Y activity and protein Y concentrations in both urine and plasma.

前言部分的写作可以采用先搭架子，再添加材料的模式。架子由三部分组成，

第一部分描述课题背景，第二部分写为什么需要进一步研究，第三部分写做了什么。论文写作先写实验结果后写前言的好处是，前言可以依照结果部分组织，达到前后对应，连贯一致的目的。

英文写作中可以简单地按照如下格式：

Starts with background description, followed by "××××　remains unknown", "××××　is less studied", "there is a need to ××××". Therefore, "we investigated ××××", "we found that ××××" and "our results provides insights (unique method) to ××××".

下面通过剖析一些论文的前言部分，介绍如何组织三个方面，写出流畅的前言。

例1.

Six Cloned Calves Produced from Adult Fibroblast Cells after Long-term Culture

Introduction

Genetic manipulation of mouse embryonic stem cells has revolutionized mouse genetic research. However, embryonic stem cells are not available in other species. Fortunately, animal cloning using cultured somatic cells offers the possibility of targeted genetic manipulations like those performed in the mouse, those somatic cells remain competent for cloning after prolonged culture. Live clones have been obtained from adult somatic cells in sheep (1), mice (2), and cows (3, 4). Furthermore, transgenic animals have been produced by cloning gene-transfected fetal somatic donor cells (5, 6) (开始先介绍动物克隆的现状).

However, to date, successful somatic cell cloning has been largely limited to the use of the donor cells either fresh (2) or after short-term (under 10 passages) in vitro culture (1, 3–6), which would not allow targeted gene manipulations (然后用however这个转折词，写出现在的不足之处).

This study was conducted to test the cloning competence of skin fibroblast cells after prolonged in vitro culture, using an aged (17-year-old) elite bull. In this paper, we report that normal live clones were produced from cultured adult somatic cells in a cattle model after up to 3 months of culture (passage 15). Our finding offers promise for producing site-specific genetically modified animals such as gene knockout animals by somatic cell cloning (第三段写我们做了什么，发现了什么和科研的意义).

例2.

Quantification and Structural Identification of Related Phenolic Compounds in the Raw Medicinal Material Honokiol

Introduction

Honokiol, whose chemical name is 3',5-di-2-propen-l-yl-[1,1'-Biphenyl]-2,4'-diol, is an effective component in the stem bark of *Magnolia officinalis,* which has been used as a traditional Chinese medicine (TCM) for treating retention of dampness in the middle-jiao marked by abdominal distention, nausea, vomiting, stagnation and constipation of indigested food and qi, cough and asthma due to accumulation of phlegm in the lung. Raw medicinal material honokiol was isolated from the stem bark of *Magnolia officinalis* Rehd. et Wils. As previously reported, honokiol possess a variety of pharmacological activities, such as antibacterial[1-4], anti-inflammatory[5], antioxidation[6-12], antiplatelet[13,14], antiarrhythmic[15], antianxiety[16,17], antitumor[18,19] and neurotrophic effects[20], underlying the importance of establishing its quality standard (前言先介绍中药 honokiol 的药性和有效成分，字数和内容都合适).

In this paper, for quantitative analyses of the related phenolic compounds in raw medicinal material honokiol, we isolated identified seven neolignans from the remaining substance produced during the preparation of the raw medicinal material honokiol *by silica gel column, semi-preparative HPLC or both. Their structures were characterized by analysis of 1D and 2D NMR spectra and their molecular formulas were determined by high resolution ion-trap time-of-flight mass spectrometry (HR-IT-TOF-MS). In the seven compounds, magnaldehyde E (4), magnaldehyde B (6) and 8',9'-dihydroxyhonokiol (7) were the major components, whose content was significantly higher than that of other four compounds (see* **Fig.** *2). 7-O-ethylhonokitriol (3) and 8', 9'-dihydroxyhonokiol (7) were the oxidized derivatives of honokiol*[21] *and both of them were novel compounds. The erythro-7-O-methylhonokitriol (1) and the threo-methylhonokitriol (2) have common planar structures and their relative configurations were firstly identified.* In addition, for the first time, we simultaneously determined the contents of magnaldehyde B, magnaldehyde E and 8',9'-dihydroxyhonokiol by HPLC in the raw medicinal material honokiol prepared using optimized method. The HPLC method is simple, rapid and accurate, which provided a standard method for the quality evaluation of raw medicinal material honokiol (这段应该写如何做研究，发现了什么和科研的意义。对实验结果描述太多，斜体字部分可以删除。另外，两段中间应加一小段，阐述为什么要做这个研究，比如对中药的有效成分还不够了解).

例 3.

Bioequivalence and Safety Study of Letrozole Tablet in Healthy Chinese Postmenopausal Women Volunteers

Introduction

Letrozole, systematically named 4,4′-[1*H*-1,2,4-triazol-1-yl-methylene] bis-benzonitrile (CAS 112809-51-5), is a third generation, competitive and highly specific non-steroidal inhibitor of the aromatase enzyme system, which could block the conversion of androgens to estrogens in all tissues[1]. Suppression of estrogen synthesis by inhibition of the aromatase enzyme system has been regarded as a crucial point in the treatment of estrogen-dependent breast cancers. Letrozole has been indicated for the hormonal treatment of advanced breast cancer in women with natural or artificially induced postmenopausal status, who have disease progression following antiestrogen therapy. Letrozole has been demonstrated to be associated with better estrogen suppression effect compared with other aromatase inhibitors, such as anastrozole, exemestane, formestane, and aminoglutethimide[2,3] (先文献综述药物 Letrozole 的特性).

Letrozole is rapidly and completely absorbed from the gastrointestinal tract (mean absolute bioavailability is 99.9%)[4]. Food slightly decreases the rate of absorption (median T_{max} 1 h fasted versus 2 h fed; and mean C_{max}:129± 20.3 nmol/L fasted versus 98.7 ± 18.6 nmol/L fed), but the extent of absorption (AUC) is not changed. Plasma protein binding of letrozole is approximately 60%. The concentration of letrozole in erythrocytes is about 80% of that in plasma. The apparent volume of distribution at steady state is about 1.87 ± 0.47 L/kg. Metabolic clearance to a pharmacologically inactive carbinol metabolite is the major elimination pathway of letrozole (CL_m =2.1 L/h), which is relatively slow compared to hepatic blood flow (about 90 L/h) (再总结药物 Letrozole 的药代动力学特性，这两段背景描述都合适，字数也合理).

Although the pharmacokinetics (PK) or bioavailability of letrozole has been previously examined in different populations[5-8], including healthy Chinese men, no data are available in healthy Chinese postmenopausal women subjects. (这一句交代已有知识的不足和现研究的必要).

The present study was designed to investigate the bioequivalence and safety properties of two formulations of letrozole 2.5 mg tablet, including a newly developed generic formulation (test) and a branded formulation (reference) in healthy Chinese postmenopausal women population. We established a validated LC-MS/MS method for the determination and quantification of letrozole in human plasma. Our study was necessary before the marketing of this newly developed generic formulation in China (这一段交代本论文做了什么工作及意义).

例 4.

Post-Paleozoic Crinoid Radiation in Response to Benthic Predation Preceded the Mesozoic Marine Revolution

Introduction

Predator–prey interactions may represent a significant driving force of evolutionary change[1-4], but predation and its consequences are often difficult to assess in recent communities and even more so in the fossil record. Data on fossil and extant crinoids, commonly known as sea lilies and feather stars (Echinodermata), indicate that they suffer from predation by fishes, and numerous evolutionary trends have been ascribed to such interactions[5-8]. Among these are (Ⅰ) crawling and swimming abilities in comatulids[9], (Ⅱ) choice of semicryptic habits and nocturnal–diurnal behavior among comatulids[10], (Ⅲ) increasing plate thickness and spinosity among Paleozoic crinoids[11], (Ⅳ) offshore displacement of late Mesozoic/Cenozoic stalked crinoids[12], and (Ⅴ) origin of autotomy (shedding) planes in the stalk and arms[13]. Some of these trends have served as examples of dramatic change in marine ecosystems, such as the Mesozoic marine revolution (MMR)[2,14] and the middle-Paleozoic marine revolution[15-16] (这段前言先总述 predator–prey interactions 在生物进化中的作用，特别是在 crinoids 进化中，predation by fishes 的作用).

Although predation by fish has received the most attention, crinoids may be the prey of other organisms, most notably benthic invertebrates. Until recently, few data hinted at the importance of benthic predators to crinoids, including a swimming response in a comatulid when perturbed by the predatory sea star *Pycnopodia helianthoides*[17], the presence of crinoid pinnulars in the gut of the goniasterid *Plinthaster dentatus*[18], and a crinoid arm observed in the claw of the crab *Oregonia gracilis*[19] (这段介绍最近发现的 importance of benthic predators to crinoids).

To further explore the interaction between extant cidaroids and crinoids, to test for evidence of the interaction in the geologic past, and to identify its evolutionary consequences, we conducted aquarium experiments, analyzed samples of Triassic fossil crinoids, and examined the evolutionary history of crinoids and echinoids (最后一段写科研目的和此论文做了什么，简单明了).

例 5.

Iron Enrichment Stimulates Toxic Diatom Production in High-nitrate, Low-chlorophyll Areas

Introduction

Mesoscale iron fertilization experiments have been performed in all of the major high-nitrate, low-chlorophyll (HNLC) regions[1-3] (Table 1), and the biological and chemical consequences of the resultant diatom blooms have substantially improved our understanding of marine biogeochemical cycles, including their complexities and links to climate processes. The success of these scientific efforts has led to policy proposals and emerging commercial endeavors that enhance atmospheric carbon sequestration in the deep ocean using iron-induced diatom blooms with high sinking potential. Debate on the wisdom of these ecological manipulations has focused mainly on the magnitude and quantification of stimulated carbon export[1,4] and the resultant effects on deepwater chemistry and biological communities[5]. Mesoscale iron enrichment experiments have focused on studying the broader issue of carbon cycling, rather than assessing the potential ecological impacts of larger-scale and longer-term geoengineering-designed fertilizations (第一段先写 iron fertilization 的现状和主要研究方向，用了 5 句话就表达得很清楚).

However, there has been little discussion whether the iron-enhanced diatom community composition itself may have unintended consequences to surface-dwelling organisms (话题一转，把未解决的问题提出来).

We present here results from in situ measurements and shipboard culture experiments demonstrating that oceanic *Pseudonitzschia* species produce DA and retain that capacity upon iron and copper amendment. The findings demonstrate that toxin production can occur with iron fertilization of HNLC waters, that the specific composition of commercial iron substrates is a critical parameter in the degree of toxin production, and that the total toxin production potentially could reach ecologically harmful levels during large-scale iron fertilization programs (第三段把论文的实验，发现和意义用三句话表达清楚).

例 6.

Experimental Investigation into Response of Circular Plates Subjected to Hydrodynamic Shock

Introduction

Solid phase-based forming has attracted a great amount of interest as a new technology of forming since it can compensate for the disadvantages of the conventional forming processes. However, it still has difficulties in industrial production due to problems such as the reheating of the billet, high manufacturing costs, and an inability to produce large parts[1]. In sheet metal forming, the amount of deformation has always attracted immensely experimental and analytical research efforts (先简介课题领域的现状和问题，此后作者还举了很多例子来充实现在的问题，在此删略).

The purpose of the present study is to gain further understanding of the metal plate behavior subjected to hydrodynamic impact loading by testing ferrous and nonferrous metal plates, including the effect of plate thickness and the transfer energy (最后写出此论文的科研目的和方向).

例 7.

Development of a HPLC Method and Simultaneous Quantification of Four Free Flavonoids from *Dracocephalum Heterophyllum* Benth.

Introduction

Medicinal plants have been employed in various traditional medicines throughout the world since ancient time. They have been a rich source of chemicals and thus many bioactive compounds have been isolated in their pure form[1]. Flavonoids are a class of naturally occurring plant secondary metabolites imparting protection to the reservoir[1,2]. They are compounds of low molecular weight and are chemically polyphenolic in nature presenting a common benzo-γ-pyrone structure[3]. They have enormous biological and pharmacological activities conferring many health benefits to the human[1,2].

Dracocephalum heterophyllum Benth. is a small perennial aromatic herb belonging to the family Lamiaceae and has been of medicinal importance in Chinese traditional medicine. It is used in traditional way of treatment of tracheitis and cardiovascular disease in Xinjiang and in Tibet region of China[4-6]. The essential oil of the plants has been shown to possess antimicrobial and antioxidant activities and thus can be used in cosmetics, food, and pharmaceutical industries[8]. Aerial part of *Dracocephalum heterophyllum* is said to contain as many as 10 types of flavonoids, and among these are luteolin, kaempferol, diosmetin, and chrysosplenetin.

HPLC is the method of choice among the chromatographic techniques for the analysis of flavonoids which needs no derivatization and thus reduces the time consumption in comparison to GC[1,2,10]. Moreover, it is safe for flavonoids as it can be operated even at room temperature thus avoiding the risk of decomposition of compounds like flavonoids at high temperature (此论文是关于用 HPLC 分析中草药 flavonoids 有效成分的研究，作者在前言部分先总述中草药的广泛使用，flavonoid 的有效成分，再介绍特定的草药 *Dracocephalum heterophyllum* Benth.，再介绍 HPLC 分析。介绍了三个方面后，才回到论文的研究目的。背景介绍有些长，也可以一句话带过中草药的广泛使用，直接综述本研究的草药).

The present paper deals with the quantification of four flavonoids, namely, luteolin, kaempferol, diosmetin, and chrysosplenetin, which are available in free aglycon forms in the plant under study, using the HPLC in a single run. This is the first paper describing the quantification of these four flavonoids in the plant extract under study (最后一段写此论文做了什么和意义).

例 8.

Effects of High Pressure on the Accumulation of Trehalose and Glutathione in the *Saccharomyces Cerevisiae* Cells

Introduction

Environmental stresses cause increased energy expenditure in yeasts, which induce changes in the metabolism of yeast cells and accumulation of some protector molecules[1]. Trehalose and glutathione (GSH) are examples of these protector molecules produced by *Saccharomyces cerevisiae* (*S. cerevisiae*) under stress conditions, such as salt shock, nutrient depletion, osmotic shock, and heat increase. High pressure has been widely used in medical instrument hygiene and food research in the last decade. High pressure is typically viewed as a denaturing factor[2-4] and has thus been used in food processing and sterilization of medical apparatus. However, there is little report about the effects of high pressure on the growth and metabolism of microbe cells (前言一开始就综述 stresses 对生物的影响，并引出需要研究的未知问题，即 pressure 对生物的影响).

Trehalose (α-D-glucopyranosyl-α-D-glucopyranoside) is a non-reducing disaccharide of glucose found in bacteria, fungi, plants and insects[5,6]. It is nontoxic and biologically inert. It also has protective effects on the properties of proteins, cell membrane and other active substances, which make it useful as a preservative in foods, pharmaceutical products and cosmetics. The tripeptide glutathione (GSH; 1-γ-glutamyl-cysteingl-glycine) is the major non-protein thiol compound in yeast. It has a wide range of functions, such as detoxification of reactive metabolites and free radical scavenging[7,8].

As stress protector molecules in yeasts, trehalose and glutathione are synthesized and accumulated when microbe cells encounter extreme conditions, such as high temperature, abnormal osmotic pressure, and the presence of heavy metals and toxic reagents[9-12]. Taking advantage of this property, trehalose and glutathione are in fact produced by the fermentation of *S. cerevisiae* cells under stress conditions[13-15]. Therefore, understanding the mechanisms of trehalose and GSH formation and accumulation under various forms of stress has become a major research topic in baker's yeast industry, as well as in wine making and brewing industries (作者用两段分别简介两个研究的对象 Trehalose 和 GSH，及它们在生物中的变化).

In this report, we investigated the effects of high pressure on the accumulation of two stress protector compounds, trehalose and GSH, in *S. cerevisiae* cells (最后一句交代作者做了什么研究).

例 9.

Preparation and Property of mPEG-PLA/pluronic Mixed Micelles and Their Role in Solubilization of Propofol

Introduction

Propofol, chemically named 2, 6-diisopropylphenol, is a highly effective and rapid intravenous anesthetics in clinic. The greatest advantage for propofol is the rapid recovery even after long periods of anesthesia. However, some drawbacks, such as poor water miscibility (150 μg/L)[1] and high lipophilicity (log$P$=4.16), were present[2], which promote to develop the alternative formulations such as oil/water lipid emulsion containing phosphatidyl choline, soya bean oil and glycerol (Diprivan®, Zeneca UK)[3], microemulsion[4], inclusion complex[5] and polymeric micelle[6,7] in order to improve drug solubility. Unfortunately, lipid-based emulsions and unique commercial preparation suffered from several limitations including poor physical stability, the potential for embolism, and rapid growth of microorganisms[8,9]. Therefore, there is an urgent need to develop an ideal formulation to solubilize propofol efficiently and solve some of the aforementioned problems for lipid-based emulsion (第一段先简介药物 Propofol 的优点和缺陷及急需研发新剂型的必要，也就是为什么要做这个研究).

In the past few years, polymeric micelles formed by amphiphilic copolymers have been extensively studied for their prominent superiorities among the emerging nanoscale carrier systems. They provided several advantages including hydrophobic drug solubilization, controlled drug release, escaping from reticuloendothelial system (RES) uptake, and efficient accumulation in pathological tissues with permeabilized vasculature such as tumor vial the enhanced permeability and retention (EPR) effect[10]. More recently, a large number of studies on mixed polymeric micelles have appeared because the prominent advantages for different types of copolymers concentrated on a single polymeric micellar system. Di/multifunctional copolymer micelles can be realized by preparing mixed micelles[11,12]. The loading content and stability of drug in mixed micelles can be improved greatly with different kinds of copolymers compared with single copolymer micelles[13-16]. The release and function of micelles can be modified to be desirable by forming mixed micelles[17,18] (本段简介药物剂型 polymeric micelles 的最新成果及其可以利用在改进药物 Propofol 上。前两段的写作都是很合理的，前言部分只需要写个科研成果小总结，就顺利完成).

In this study, the objective was to develop a new mixed polymeric micellar formulation comprised of methoxy poly(ethylene glycol)-poly (lactide) polymer (mPEG–PLA) and pluronic triblock polymers to enhance the solubility of propofol. *The amphiphilic block polymer mPP was synthesized and used in this study because of its excellent micelle formation, drug loading capability and release behavior. Pluronic block copolymers that consist of hydrophilic ethylene oxide (EO) and hydrophobic propylene oxide (PO) blocks in a basic EO–PO–EO pattern, were used in the study for their commercial availability, biocompatibility and safety. Thus, the mixed micelles using mPP and pluronic copolymer (P105) were prepared to solve the problem for lipid-based emulsion.* The prepared mixed micelles were characterized by particle size distribution, drug loading content, encapsulation efficiency, free drug concentration, stability, *in vitro* release profile and pharmacological effect (最后一段写做了什么和实验结果及意义。科研结果部分太长了，斜体字部分可去掉。若再加一句科研成果的意义就更好了).

例 10.

Design and Synthesis of cADPR Analogues with Simplified Ribose and Nucleobase

Introduction

Cyclic ADP-ribose (cADPR, **1**) was discovered as a Ca^{2+} mobilizing metabolite of β-nicotinamide adenine dinucleotide (NAD^+) by Lee and coworkers in sea urchin egg homogenates[1]. Since its discovery, numerous cell systems have been described to utilize the cADPR/ryanodine receptor (RyR) Ca^{2+} signaling system to control Ca^{2+}-dependent cellular responses, such as fertilization, secretion, contraction, proliferation and many more[2,3]. To elucidate the detailed molecular mechanism of cADPR/RyR signaling pathway, various cADPR analogues have been synthesized. These analogues include the modification of southern and northern riboses[4-6], purine[7-10] and pyrophosphate[11,12] of cADPR. They can agonize or antagonize cADPR/RyR calcium signaling pathway (先简介什么是 Cyclic ADP-ribose (cADPR)，它的生物学功能和为什么要研究这个化合物).

In our previous works, a cADPR analogue in which the northern ribose was substituted by an ether strand termed cyclic inosine diphosphoribose ether (cIDPRE, **2**) was synthesized; its biological activity was shown in permeabilized and intact T-Jurket cells. The cADPR analogue in which both the northern and southern ribose were substituted by ether strands (cIDPDE, **3**) was found to be the agonist of cADPR/RyR calcium signaling pathway[13]. A much more simplified cADPR analogue with ribose moieties replaced by ether or carbon strands and nucleobase replaced by triazole, (cTDPRE, **4**), could also act as membrane-permeable calcium agonist[14] (再综述过去的科研成果，特别是自己实验室的成果).

To understand the structure-activity relationship of simplified cADPR analogues better and prepare probes for the investigation of the molecular mechanism of cADPR regulated calcium pathway, we have designed and synthesized novel cADPR analogues that integrate three parts of modification, nucleobase, northern ribose and southern ribose (最后一段用两句话写科研目的和为什么要做论文研究，及此论文的实验内容，简要且明确).

1.5 Materials and Methods or Experimental Section

　　实验部分的写作要详细、具体，但不能冗长，要达到读者根据实验描述能重复出实验结果的目的。所以在生物实验中，对试剂的来源，样品的出处、保存，操作步骤等都要有详细的描述。化学实验一般不要求注明试剂来源，因为试剂来源一般不影响结果；但实验条件要详细、定量，试剂来源可以写在 Materials 部分，也可以在实验步骤中加括号注明，比如 Insulin（Sigma #I3579）。Materials and Methods 部分的写作应与 Results 部分相对应。任何一个实验结果都要有相应的实验方法。先写 Results 部分的好处之一就是 Materials and Methods 部分写起来很有针对性，有了实验结果，Materials and Methods 部分写什么就明确了。

　　实验部分的写作方法应注意以下几点：

① 用小标题来组织实验部分，按合理的顺序来讲述实验是如何操作的；

② 都用过去时，以第三人称和被动句为主，偶尔使用第一人称；

③ 特别注意实验要定量。对用量、时间、温度等都要具体写明；

④ 明确写出具体条件，试剂，溶液，所用仪器型号，生产厂家等，避免使用含糊的代词，比如用 1% ethanol，而不用 solution 1；用 1 mg/mL，而不用 indicated concentration 等；

⑤ 写明数据处理和分析方法。

实验部分从写作上来讲相对简单，一句一句的并列可以接受；但实验步骤往往是一步接一步，在前后连接上，往往会出现过多使用"Then，After"等词，造成单调冗长的感觉。为了避免出现多个重复的单句，几个单句可以合并。例如：

After 4 days of treatment the animals were pulsed with bromodeoxythymidine before sacrifice, euthanized, tumors removed, and fixed in 4% paraformaldehyde overnight and then embedded in paraffin.

20 μL samples were removed, precipitated with 60 μL acetone, centrifuged at 4 ℃ for 10 minutes at 3000 RPM, and transferred to a clean vial.

The combined organic layers were washed with water, washed with brine, dried over sodium sulfate, filtered, concentrated and purified on silica to yield 0.8 g of a white solid.

对某领域内熟悉的一些方法，提到即可，不需详细描述。比如"Solvent was removed under vacuum"至于如何去做就不必写出。再比如"Proteins were separated on a PAGE gel"即可，不必描述如何走"PAGE gel"。

下面是一些常见的 Materials and Methods 的例子。

例 1.

Experimental animals

Male Sprague-Dawley rats (weighing 220–250 g) were purchased from the Laboratory Animal Center of Northeast University of Traditional Chinese Medicine. The animals were kept in an environmentally controlled breeding room for 1 week before the experiments. The rats were fasted for 12 h with water ad libitum prior to intravenous administration of compound Y. All procedures involving animals were in accordance with the Regulations of Experimental Animal Administration issued by the Ministry of Science and Technology of People's Republic of China.

实验动物

雄性 Sprague-Dawley 大鼠（体重 220-250g）购自东北中医药大学实验动物中心。实验前，动物在环境控制的饲养室中饲养了 1 周。大鼠在静脉注射化合物

Y 前禁食 12 小时，并自由饮水。所有涉及动物的程序都符合中华人民共和国科技部颁布的《实验动物管理条例》。

例 2.

Cell lines

Mouse sarcoma S180 cell lines were provided by Dingguo Biological Technology Co., Ltd (Beijing, China). All cells were maintained in DMED supplemented with 10% FBS. Cells were incubated at 37 ℃ in 5% CO_2 atmosphere. Mice were injected intraperitoneally with 1×10^6 viable cells.

细胞系

小鼠肉瘤 S180 细胞系由 Dingguo 生物技术有限公司（中国北京）提供。所有细胞都在补充有 10% FBS 的 DMED 中维持。细胞在 37℃，5%的 CO_2 环境下培养。小鼠腹腔注射 1×10^6 个存活细胞。

例 3.

Liquid chromatography conditions

The HPLC system consisted of a pump (BiSepTM-1100 pump, Unimicro Technologies Inc., Pleasanton, CA, USA), a UV/Vis detector (BiSepTM-1100 detect, Unimicro Technologies Inc., Pleasanton, CA, USA), and a Model 100 Column Heater (CBL photoelectron technology, Tianjin, China). Separation of analytes was achieved at 20 ℃ on an UltimateTM AQ-C18 column (250 mm × 4.6 mm, 5 μm). The mobile phase (pH 5.05) was pumped at a flow rate of 1.0 mL/min and consisted of acetonitrile-water (31/69 V/V), containing 0.002 mol/L sodium dodecyl sulfate, 0.0125 mol/L sodium dihydrogen phosphate, and 0.015 mol/L triethylamine. The UV detector was set at 236 nm. The chemical structures of metformin and IS are shown in Fig. 1.

液相色谱条件

高效液相色谱系统由泵、紫外/可见光检测器和 100 型色谱柱加热器组成。分析物的分离在 20 ℃，UltimateTM AQ-C18 色谱柱上实现。流动相流速为 1.0 mL/min，由乙腈-水组成（31/69 V/V），含有 0.002 mol/L 十二烷基硫酸钠，0.0125 mol/L 磷酸二氢钠和 0.015 mol/L 三乙胺。紫外检测器被设置在 236 nm。二甲双胍和 IS 的化学结构见图 1。

例 4.

General experimental methods

All chemicals were purchased as reagent grade and used without further purification, unless otherwise noted. Dichloromethane was distilled over calcium hydride. Analytical TLC was performed on silica gel 60 F254 precoated on glass plates, with detection by fluorescence and/or by staining with 5% concentrated sulfuric acid in EtOH. Column chromatography was performed employing silica gel (230−400 mesh). ^{1}H NMR spectra were recorded on advance spectrometers at 25 ℃. Chemical shifts were referenced with tetramethylsilane (δ = 0 ppm) in deuterated chloroform and deuterated dimethyl sulfoxide (δ = 2.50 ppm). ^{13}C NMR spectra were obtained by using the same NMR spectrometers and were calibrated with CDCl3 (δ = 77.00 ppm) and DMSO (δ = 39.52 ppm). High-resolution mass spectrometry was performed on FTICR mass spectrometer.

常规的实验方法

除非另有说明，所有的化学试剂均为购买的试剂级，未进一步提纯。二氯甲烷在氢化钙上蒸馏干燥。TLC 分析是在预涂的硅胶 60 F254 玻璃板上进行的，通过荧光和/或用 5%浓硫酸在乙醇中染色来检测。柱色谱法采用硅胶（230−400 目）进行。^{1}H NMR 在 25℃的高级光谱仪上记录。化学位移是以氘代氯仿中的四甲基硅烷（δ=0 ppm）和氘代二甲亚砜（δ=2.50 ppm）为参照。^{13}C NMR 也是由相同的光谱仪测得，并用 CDCl$_3$（δ=77.00 ppm）和 DMSO（δ=39.52 ppm）作为参照。高分辨率质谱分析在 FTICR 质谱仪上进行。

1.6 Results

论文写作不妨先从简单，易写，自己熟悉的实验数据和结果开始，先完成 Results 部分。Results 部分都是分成若干个小段落，一段一段写起来比较容易。在写 Results 之前，一般要把主要结果和数据作成 Table 和 Figure，而这些 Table 和 Figure 会作为每个小段落的核心，围绕这些数据来组织。将实验数据列表格和制图表、曲线，是非常重要的，要用较长时间来精心准备。Results 写作之前还应参考有关文献，看看文献中是如何组织数据和实验结果的。总起来讲，Results 部分写起来比较容易，语言上应注意用过去时，用词上要准确。

Results 段落是客观地讲述实验结果，一般不加讨论和解释。写作方法应注意以下几点。

① 通常用过去语句，但某些结果，如计算等应用现在时态。

② 有的学科只用文字来讲述实验结果，但大部分学科往往添加 Figure 和

Table 来帮助组织数据。制备 Table 和 Figure 是 Results 部分写作的重要部分。Table 和 Figure 的数量不要太多，能用文字说明清楚的，就用文字，而不用 Table 和 Figure。

③ Figure 和 Table 要分别给予序号，不统一排序。

④ Table 要有个标题，标题就写在 Table 的上方。表格编号后应该有个 "."，标题结尾没 "."，注解文字写在 Table 的下方。Table 有时需要照相排版，所以草稿中的 Table 要单独成页。Table 应该附在稿件的后面，如放在 References 之后。例如：

Table 1.   Compound A and compound B affinity for human and mouse melanocortin receptor subtypes

Receptor	pK_i (K_i, nmol/L)	
	Compound A	Compound B
Human MC1	7.39	8.09
Human MC4[a]	6.13	6.53
Mouse MC1	9.02	5.68
Mouse MC4[a]	5.63	5.89

The affinity of ligands for melanocortin receptors in HEK cell membrane was measured by inhibition of [^{125}I]-MSH binding, as described in Methods a, CHO cell line was used.

⑤ 考虑到排版的需要，期刊要求草稿中的每个 Figure 要单独占一页，并且这一页只是 Figure，把序号写在 Figure 页的背面。Figure 应该附在稿件的后面，如放在 References 之后。例如：

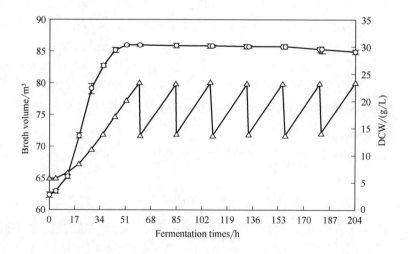

⑥ Figure 的说明（Figure Caption）要单独写，不与 Figure 放在一起，一般可以放在 References 部分之前或之后。但不同 Figure 的说明可以排在一起。Figure 说明中只写介绍 Figure 的文字，不对 Figure 的结果进行解释。例如上面 Figure 的说明：

Figure Captions

Figure 2. Time course of broth volume (△) and dry cell weight (DCW, ○) in repeated fed-batch fermentation of penicillin G. From 60 h to finish, the working volume was controlled in the range of 72−80 m^3, 8.0 m^3 of broth was drawn off at a 24 h interval with the mycelium concentration (DCW) kept stable. Data was the average of 16 batches.

⑦ 方程式一般给一个序号，写在方程式右边的括号中，例如：

The distance traveled is given by S:

$$S=vt \qquad (1)$$

where v is velocity and t is time.

Results 段落开头时可以直接讲述结果，不必加引言或过渡句。经常会用"为了×××，而做×××实验"作开头。下面是个类似的句子：

In order to study the pollution of rivers around the Boston area, water samples from different rivers were analyzed to determine their PPC concentrations. As shown in Table 1, ……（描述实验结果）

In order to estimate the effect of the active control strategy, a simple testing device was built. A schematic representation of the simulating device is shown in Fig 5.

Results 部分一般要分段，小标题写作。具体写作时可以考虑先把小标题列出来。

例 1. 对一篇化合物合成和生物活性研究的论文，可有这么几个小标题：

1. Reaction condition screenings

2. Synthesis of intermediate

3. Synthesis of final product

4. Reaction mechanism study

5. Biological activity investigation

例2. 一篇分析方法和应用的论文可以有如下几个部分：

1. Development of the method

2. HPLC conditions

3. Optimized method

4. Validation

5. Application of the method

例3. 一篇有关污染物的稿件可有如下几个小标题：

1. Seasonal variation of contaminant X in the river

2. Composition of contaminant X

3. Source identification

4. Mass flow of contaminant X

5. Hazard assessment

例4. 一篇有关疾病药物治疗的稿件可有如下几个小标题：

1. Patient population and physiological characteristics

2. The survival rate and recurrence of disease in the patient group

3. Comparison of survival between treated and placebo group

4. Analysis of risk factors

这部分写作的主要特点是要清楚、简洁、准确，避免使用含糊的字词，如：situation，above，modulate，this，the former，the later，scientific basis，value，performance 等。要用很明确的字词，如：increase，decrease，inhibit，stimulate，activate，this result，the sample preparation procedure 等。

不同课题的写作内容差距很大，在此仅举几个例子。

例5.

Effects of Pressure on the Growth of *S. cerevisiae*

The growth curves of *S. cerevisiae* cultured under different pressure of 0.1, 0.5, 1.0, and 1.5 MPa were generated. As shown in Fig. 1, yeast cells showed a pressure dependent decrease in the growth rates (酵母菌的生产曲线用图1展示，这样清晰，一目了然。随后的文字用来解释酵母在不同压力下的生长曲线，特点，比较和需要特别注意的参数). The growth at higher pressure of 0.5, 1.0 and 1.5 MPa were much slower than that under atmosphere pressure (0.1 MPa), resulting a much lower biomass. *S. cerevisiae* reached stable growth when cultured at 30 ℃ at 14 h under atmosphere pressure and maintained the stable growth until the 36 h of incubation. The growth started to decrease slightly after 26 h. However, when higher pressures were applied, the stable growth periods were shortened dramatically and the followed growth decreases were much sharper. For example, under 1.0 MPa pressure,

S. cerevisiae reached stable growth at 15 h and the growth started to decline sharply at 20 h. Even though the growth periods were much shorter under higher pressures comparing to normal pressure, the durations of stable growth periods under higher pressure of 0.5, 1.0 and 1.5 MPa were similar. Moreover, the log phase of cell growth was also delayed under higher pressures. The log phase of normal growth started at 6 h. But the log phases of higher pressure growth started at 8-10 h. In general, the higher pressure in the bioreactor resulted in deviations from the growth curve normally obtained in fermentations of *S. cerevisiae*.

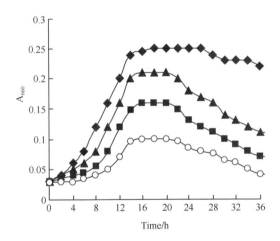

Fig. 1. The growth curves of *S. cerevisiae* cultured under different pressures.

They represented the mean of three measurements. The standard deviation of each data point was below 16%: (●) 0.1MPa; (■) 1.0MPa; (▲) 0.5MPa; (○) 1.5MPa.

例 6.

Characterization by XPS

XPS is a powerful technique that allows the detection and determination of the elemental and chemical composition of electrode surface and is an effective tool for the quantification of immobilized or adsorbed proteins. Here, we have used this technique to demonstrate that GOD was immobilized successfully onto WCNT-modified TP electrodes. The surface elemental compositions for TP, TP/MWCNT, and the TP/MWCNT/GOD electrode are summarized in Table 1（X射线光电子能谱的

分析结果用表格展示，简洁，清晰。文字部分用来描述每个样品的分析结果和特点）。

Acting as the substrate electrode, TP consists mainly of carbon (98.58%) and some other minor impurities from its synthesis. It is clear that the oxygen content increased after the deposition of the MWCNTs, which can be ascribed to the introduction of carboxylic groups during the oxidation of the MWCNTs. Concerning the immobilization of GOD, we observed significant increases in N and O as expected from the backbone amide and carboxyl groups of GOD, confirming the presence of GOD. In addition, the flavin adenine dinucleotide (FAD) prosthetic group that functions as the redox center for the electron-transfer process in GOD consists of a riboflavin group bound to the phosphate groups of ADP. Therefore, the appearance of Na and P in the elemental analysis also confirms the presence of GOD. With the increased percentages of N and O, the results from XPS clearly demonstrate that GOD was immobilized successfully on the TP surface, and this conclusion was further confirmed by the FTIR spectra.

Table 1 – The atomic composition of TP, TP/MWCNT, and the TP/MWCNT/GOD-electrode surfaces.							
Electrodes	F 1s (%)	O 1s (%)	N 1s (%)	C 1s (%)	Si 2p (%)	Na 1s (%)	P 2p (%)
TP	NA	1.16	0.13	98.58	0.13	NA	NA
TP/MWCNT	1.09	7.96	4.53	86.23	0.19	NA	NA
TP/MWCNT/GOD	2	13.2	6.6	75.2	0.34	1.75	0.92

例 7.

Analysis of the Samples

Fifteen batches of *Siphonostegiae Herba* samples purchased from different regions were analyzed, and the results are shown in Table 3. The content of acteoside varied from 0.564 to 7.036 mg/g, while the content of luteolin ranged from 0.641 to 3.047 mg/g. As shown in Table 3, the amounts of these two compounds differed greatly in the samples from different sources, which could be caused by the different growth environment, the age and harvesting times. These results indicated that it would be necessary to control the quality of *Siphonostegiae Herba* to ensure its therapeutic effects in clinical use. Our present study suggests that setting a minimum limit on luteolin and acteoside could be helpful to ensure the quality and clinical efficacy of *Siphonostegiae Herba*（不同样品的有效成分含量用表格展示，文字部分不需要描述每个样品的含量，只重点指出它们的变化）。

Table 3. Contents of luteolin and acteoside in Siphonostegiae Herba

No.	Source	Luteolin (mg/g)	Acteoside (mg/g)
1	Bozhou, Anhui	0.83	1.20
2	Nanjing, Jiangsu	1.43	2.15
3	Wuhan, Hubei	1.99	3.39
4	Changsha, Hunan	1.00	0.86
5	Beijing	1.60	1.67
6	Beijing	0.78	1.50
7	Anguo, Hebei	0.96	0.56
8	Shiiazhuang, Hebei	3.05	7.04
9	Lanzhou, Gansu	2.10	4.85
10	Jinan, Shandong	1.18	2.73

1.7 Discussion

许多期刊把 Results 和 Discussion 放在一起以便讨论结果的意义。Discussion 部分与 Introduction 是紧密相关的。讨论要起始于某个课题的已知情况，然后再讨论实验结果对课题研究的进一步认识和意义，当然也要说明结论存在的问题和局限等。可以说这部分是一篇文章中最难写的，如何解释实验结果需要有一定的知识基础和科研思维能力。要写好自己的实验结果及其意义和它对某个课题研究的更深入认识，需要对这个领域的过去和现状有基本了解，一般作者做了些实验，有些新数据，是知识的积累，往往不是创新，在讲述这些结果的意义时，要做到合理、合适，不牵强附会。科学结果往往是相对的，没有绝对正确，写起来要留有余地，所以会经常使用 suggest, imply, appear 等词。除了掌握写作技巧外，作者应熟悉自己科研领域的文献，善于把自己的结果与文献知识联系起来，这样才能在已有的知识背景下讨论自己实验结果的含义和意义。

讨论部分写作一般应注意以下几个问题：

① 尽量使用主动句，可以用第一人称。语句要简练，避免写冗长的句子；

② 对实验结果按其顺序进行讨论,讲述每个结果表示什么和对某个问题认知的意义，这时需要引用文献结果来支持自己的结论；

③ 讨论部分不能引入新的数据，所有数据都应来自 Results 部分；

④ 要引用已发表的他人的结果或自己以前的结果来支持自己的结论,有时也需要表明自己的结果支持他人的发现,应该写出自己的结果与已发表的文献的相同和不同之处;

⑤ 应该用一小段来阐述在你的结果的基础上,新的问题或假想是什么?你是如何去研究新的问题或假想的。

论文,顾名思义,就是要"论"或"Discuss"。通过论述,把自己科研成果和理论的重要性、意义、应用、缺陷等表述清楚,以达到交流的目的。如果把科研成果比喻成一件艺术品,那论述就是你对这件艺术品的讲解。艺术品的创造、特点、与其他类似作品的显著区别及不足等都可能是你讲解的内容。同样,论述部分一般或多或少要涉及四个方面:论文的新颖发现;与已发表成果的不同或对已有成果的补充;科研成果的意义和应用性;有待改进之处和未来科研方向。

讨论部分最容易被写成文献综述或试验结果的另一种描述。为避免这种情况的发生,讨论部分要先从自己的成果写起,讨论你的结果对课题研究的认识和意义。讨论是在论述文献的背景下,与文献知识相比,阐述论文的结果。从很多方面可以找出论文研究的新颖之处。比如:

新的化学反应或反应条件;

更简洁有效的制备方法;

新活性化合物;

新材料或材料新特性;

新工艺和高产率;

改进的分析方法;

高灵敏度,多化合物分析;

某个特殊物种的研究;

基因和酶的研究;

中国人群的药物代谢和药效;

某种疾病群体的药物代谢和治疗;

新病例的诊断和治疗;

针对中国病人群体的研究。

下面讲解一些讨论部分写作的例子。

例 1.

Discussion

In the present study, we investigated the possibility of producing normal, cloned cattle by using adult somatic donor cells after long-term culture. Our results demonstrated that long-term culture (up to 15 passages or estimated over 45 cell doublings) of skin fibroblast cells derived from an aged bull did not seem to compromise their cloning competence (一开始就简明地写出此论文做了什么和主要发现，以此来展开讨论). This finding is significant because it offers the possibility of using gene-targeted somatic donor cells for cloning. Currently, we have cultured adult cattle skin fibroblast cells for 30 passages, and the cloning competence of these cells is being evaluated. To our knowledge, these were the first normal clones born by using adult skin fibroblast cells after long-term culture (接下来鲜明地写出此科研结果的意义，指出这是第一次此类的报道). Somatic donor cells after short term culture (under 10 passages) have been used previously for nuclear transfer (1, 3, 4, 17, 18). However, the cell passage effect on the outcome of cloning cannot be concluded from those studies because they were conducted by different groups using different cell types and passage numbers, different protocols of cell preparation, nuclear transfer, and activation, and different culture systems. In the present study, we used the same standardized procedures in our nuclear transfer experiments, and all micromanipulations were performed by the same skilled microinjectionist. This particular setup makes direct comparison of cell passage effects more meaningful (接下来写与以往发表的文献相比，此论文的优势和特点。此论文的结果可以用来做 cell passage effects 的比较).

To date, the overall cloning efficiency using somatic cells has been low, with the reported efficiency ranging from 0 to nearly 10%. This could be partially explained by the high embryonic loss during the first half of gestation. Our observation of a high rate of embryonic loss between days 60 and 120 of gestation is consistent with previous reports for fetal fibroblast clones (5, 6, 17, 22) and adult cumulus cell-derived clones (4). Although low embryonic loss and high calving rates were reported in a previous study (3) using oviductal or cumulus cells for cloning, a high neonatal mortality (50%) was noted. The exact mechanisms of early or late embryonic loss and neonatal death of clones are still not clear; however, incomplete reprogramming of the donor cell genome in the current cloning scheme may be partially responsible. Abnormal placenta development for the cloned fetus has recently been reported (23). To improve the cloning efficiency, the exact mechanism for embryonic loss and high neonatal mortality needs to be investigated systematically (论文成果部分的讨论是最重要的，同时也要论述实验方法和现象的改进，不同或相同之处，这类信息对读者也很有益处。这一段就用来讨论一个实验现象 (embryonic loss)，作者也发现了与文献类似的现象，作者试图予以解释和指出改进的方法)。In this study, we demonstrated that adult somatic cells remained totipotent for cloning after long-term culture. This suggests the feasibility of targeted genetic manipulations such as gene knockout using cultured somatic cells before cloning to produce knockouts or other types of genetically engineered cloned animals. Cloning using site-specific genetically manipulated cells would be a valuable tool with applications in agriculture, medicine, and basic biological research (作者最后用三句话对论文的目的、主要发现和意义进行总结，以此结尾)。

例 2.

Discussion

Traditional Chinese medicine has been used in a range of medical practices and health interventions in China for thousands of years. Natural products screening becomes an important component of exploring novel compounds. Today natural products have been widely used for the treatment in a variety of diseases, including cancer. For anti-cancer treatment, even more than 60% of the approved and pre-new drug application candidates are natural products or synthetic molecules based upon the natural product molecular skeletons[26]. Compared to a variety of side effects of the current anti-tumor treatments, advantages of natural products include higher anti-tumor effect and lower toxicity. Furthermore, some extracts from natural products, such as *Piper longum Linn*[27], *Cassia fistula*[28], *Phellinus rimosus*[29] and *Soamsan*[30] could improve the immune function, which may contribute to boost the immune system's response to cancer. Recently, the anti-tumor effect and potential immune responses of terpenoids, such as terpenoid indole alkaloids[31], betulinic acid[32] and ceanothic acid and non-glycosidic iridoids[33] have been repeatedly reported. The unsaturated bonds acting as a Michael acceptor in biological systems could be a possible mechanism underlying the anti-tumor activity. Therefore, terpenoids may preferably react with nucleophiles, especially thiol groups of proteins[34]. We have demonstrated that AE has a certain inhibitory effect on the proliferation of the small blood vessels in the chick chorioallantoic membrane experiments, indicating that the inhibition of tumor angiogenesis may serve as another mechanism for its anti-tumor effect (此论文是关于有效天然产物的分离和活性的，作者先论述天然产物的优点，再描述某一类天然产物的活性和现状，再转到此论文的科研。这段论述很类似前言部分，写得更像

Introduction。比较好的写作应该是简短介绍一些文献研究，而重点指出新化合物 AE 的发现和活性研究)。

Based on the spleen index (SI), thymus index (TI) and parameters of hematological, AE could exhibit anti-tumor effect without affecting the immune system in STB mice, which is one of the most important advantages of natural products. Spleen and thymus, two important components of the immune system, can generate a large number of immune cells, such as T cells and B cells. Normal immune system could activate cytotoxic effector cells, such as T cells and NK cells[35], which may fight again tumor with cytotoxic effector cells. The parameters of hematological also reflected the situation of immune system. In STB mice treated with AE, the SI, TI and parameters of hematological did not decrease significantly at dose of 5 or 10 mg/g body weight, and there was a modest increase at the dose of 10 mg/kg body weight, suggesting that AE did not affect the function in immune system, or even slightly enhance the immune system at doses with significant anti-tumor activities. In addition, we also observed increase of life span in AE-treated STB mice (作者针对 AE 化合物的一个特性 without affecting the immune system 进行阐述和讨论，这是区别于其他化合物的地方，这样的讨论是合理的)。

1.8 Conclusion

有的期刊要求单独写一段 Conclusion，即使没有这个要求，一般在 Discussion 的最后都加一段总结。把论文中的最主要结果简洁地叙述一遍，写作要求与 Abstract 部分类似，但要避免和 Abstract 用重复的句子。

1.9 Acknowledgments

Acknowledgments 写在 Discussion 和 References 之间，一般是感谢项目资助单位和对论文的实验和写作部分有贡献但又不是作者的同事。

例如：

Acknowledgments. This work was supported by a grant from National Institute of Health (GM 3478672). The authors would like to thank Wei Wang for excellent technical support and Professor Liyi Zhang for critically reviewing the manuscript.

越来越多的期刊要求提供每个作者的贡献（credit authorship contribution statement），利益冲突说明（declaration of competing interest）等。建议参考期刊上已发表的论文，按照同样的格式写作。

1.10 References

References 要单独另起一页。

索引文献的写法，不同期刊的要求差别很大，作者应严格按照所投期刊的要求去撰写。用 Endnote 软件来帮助组织文献是值得提倡的，特别是在需要索引大量文献时。Endnote 还可以按所投期刊的要求来编排。不过用 Endnote 编辑后的文献，还要再检查一遍，由于一些特殊原因，常常会出现页码、作者信息等小的错误。

文献在论文中的引用可以用三种方法标出。

（1）Cited in parenthesis or bracket, like (1) or [1]

The method was developed by Smith's group (1).

The method was developed by Smith's group [1].

When cited at the end of a sentence, they are written before "," or ".".

（2）Cited by number in upper corner

The method was developed by Smith's group. [1]

The method was developed by Smith's group[1].

When cited at the end of a sentence, sometimes it is after "," or ".", sometimes it is before "," or ".".

（3）Cited by author names

① Single author: the author's name (without initials, unless there is ambiguity) and the year of publication, like (Zhang, 1998), (Allen, 2001；Seymour, 2003).

② Two authors: both authors' names and the year of publication, like (Allen and Wang, 1998).

③ Three or more authors: first author's name followed by "et al." and the year of publication, like (Gao et al., 2007). There is a "." after et al.

Citations may be made directly (or parenthetically).

Kramer et al. (2000) have recently shown ⋯.

Groups of references should be listed first alphabetically, then chronologically. Examples: "as demonstrated (Allan, 1996a, 1996b, 1999；Allan and Jones, 1995)".

More than one reference from the same author(s) in the same year must be identified by the letters "a", "b", "c", etc., placed after the year of publication.

每个期刊对文献的书写方法都有自己的要求，请查看期刊的"Guide for Authors"。

一般要注意的是：

① 排序可按出现先后，也可按作者姓名字母排列；

② 作者姓名的写法,可先写姓也可后写姓,比如Smith A. J. 也可写成A. J. Smith, 但全文要统一；

③ 注意期刊名称用意大利化字体还是要用黑体,应特别注意期刊名称的简写方式，不能随便臆造（见附录）；

④ 发表年、卷、页数的写法和排列方式。

1.11 修改稿件

稿件需要修改多次才能投稿，修改十几遍也是很正常的。一篇没有经过认真准备的稿件，肯定会被要求返修，甚至还会让审稿人对论文研究的质量也产生怀疑。所以只有当自己完全满意后才可以考虑投稿。修改稿件时，应注意以下几个方面。

（1）认真阅读期刊的"Guide for Authors"，检查稿件是否符合要求

（2）稿件段落

应该按以下顺序排列：

Title page (one page)

Abstract and Keywords (one page)

Main text：Introduction, Experimental Section, Results, Discussion

Acknowledgments

References

Figure legends

Tables

Figures

（3）Title page (标题主页)

作者名称，名字在前，姓氏在后。比如 Xiaodong Wang。

邮编和省市之间不用逗号，只用空格。比如"Beijing 100082"而不是"Beijing, 100082"。

"Running title"就是标题的简缩版。 比如 "Characterization of two forms of recombinant insulin-like growth factor-1：activities and interactions with insulin-like growth factor-1 binding proteins"，

"Running title"可以是 "Characterization of insulin-like growth factor-1"。

（4）Abstract （摘要）

摘要的字数、格式是否符合期刊要求。

（5）文献引用格式

用作者名称还是数字排号，如果用数字排号，又是什么格式：(5) 或 [5] 或 [5]。

（6）文献书写格式

查看每篇文献是否按"Guide for Authors"的要求书写。

（7）检查一致性

经常出现的不一致写法：

Figure 全拼 或 Fig. 缩写。

Table 1. A list of…(1 后加点) 或 Table 2 The binding activities of (2 后没加点)

Zhang et al 或 Zhang et al. 或 Zhang *et al.*

J. Biochem. 或 J Biochem (J 后没加点)

小标题是否按顺序依次排列。

检查文献列表是否都被引用和引用的文献是否都被列表。

（8）主语和谓语的单复数是否一致

不一致的情况经常发生，需要逐句检查。

(9) 句子的时态用法是否一致

在引述文献结果时，过去发表的结论可以认为是已经被承认的事实，应该用现在时。讲述实验和实验结果用过去时。

讲述 table 或 figure 中显示的结果时，可以用现在时。计算的结果和统计分析结果应该用现在时。

(10) 检查冠词 a、an、the 的使用

最常见的是漏掉，再就是定冠词 the 用多了。

第2章
稿件投稿和发表

2.1 期刊选择和稿件投稿

对如何选择投稿期刊，可以考虑这几个方面。

（1）查看稿件索引的文献是在什么期刊上发表的，这也是稿件科研领域内的常见期刊。

（2）查看一些感兴趣的期刊的中科院分区和影响因子（IF），对自己稿件的水平能达到哪个分区以及多高的影响因子要心中有数。

（3）浏览期刊的网页，阅读近期的论文摘要，看有没有自己稿件类似内容的论文被接受，也就是看稿件的科研方向是否与期刊的内容一致。

（4）看有没有国内和其他非西方国家作者的稿件发表。

经过这样的调查研究后，选择 2~3 个可以考虑的期刊。

投稿只是发表 SCI 论文的第一步。如何应对投稿后期刊编辑和审稿人可能提出的问题，对稿件能否被接受起重要作用。现在，大都实行同行评议，一个稿件不需修改就被接受的概率很小，几乎都需要按编辑或审稿人的要求修改后才会被接受。

当期刊主编收到稿件后，先编号入档。这时作者会收到期刊社来的编号信。主编的任务是浏览稿件的内容，决定稿件是否属于期刊所涵盖的专业领域，再就是决定成果的重要性和新颖性是否达到期刊的要求。如果主编认为稿件的内容或重要性不适合在其期刊上发表而予以退稿的话，他的决定一般是不会改变的。特别是对一些一流期刊，退稿是正常情况，并不一定表明稿件不好，毕竟一个期刊能刊登的论文是有限的。像《Nature》和《Science》这样要求很高的期刊，只有很少的稿件能被接受，大部分被退稿。

当期刊编辑认为稿件适合发表时，会将稿件寄给 2~3 个审稿人评审（也有可能会是 4~5 个审稿人）。期刊社有自己的审稿人名单，当一个人在某个领域新近发表一些论文后，自然而然会有期刊编辑邀请审稿。另外，期刊也会让作者推荐审稿人，有的期刊没有特别要求，有的则要求推荐非本国的审稿人。审稿人都是在相关领域做研究的科学家，审稿没有报酬，还花不少时间，但它是科学工作者的义务，也是一种荣誉，大多数人都严肃对待，做好评审工作。期刊社一般要求审稿人在 1~3 周左右把对稿件的评语返回编辑部，编辑根据审稿人的意见，同时也考虑自己的看法，决定是否接受稿件。不要忘了编辑也是审稿人之一，当审稿人之间的意见不同时，编辑就起到仲裁人的作用。这时有三种可能的结果：

① 接受，但需要做一些小的改动。
② 需要较大的改动，如补充数据或修改结论等，经过修改后，再次投稿。
③ 不能被接受。

最理想的是第一种情况，按编辑或审稿人的要求做一些修改就可以寄回编辑部发表。第二种情况下，审稿人对稿件的数据或结论与作者有不同的看法，可能会涉及一些核心问题，如数据不全或结论不合理等。这时最好的办法是按照编辑的要求去做，尽量满足审稿人的合理要求。如果审稿人的要求不合理，应向编辑详细论述自己的观点。有时审稿人也有误解的时候，要求也许有过分之处。只要做了最大努力去修改，编辑会综合各方面的情况，最终接受修改的稿件的概率还是很大的。有的作者不愿意再花时间去修改，而转投要求低一些的期刊，这样就失去了在较好期刊上发表成果的机会。

出现第三种情况时也不能说就完全没有希望了。还要看问题出在哪里。作者应仔细阅读审稿人的评语（comments）。如果审稿人的评价合理，稿件的确有严重不足之处，那就没有反驳的余地。如果评语有不合理之处，或审稿人误解了，这时应该向编辑做详细说明，编辑还是考虑合理的反驳的。除此之外，如果遇到一个审稿人不同意发表，其他审稿人和主编觉得稿件还可以，会通知作者，认真修改后，以新稿件再投上去。如果已经明显被拒绝，可以考虑重新选择期刊，再进行一轮新的投稿。

期刊主编的一边是投稿者，另一边是期刊社，主编对两边都要照顾到。编辑希望其期刊能吸引高水平的稿件，愿意与作者们合作，发表尽量高水平的论文。只要稿件的内容合适，且具有一定的重要性和新颖性，编辑还是有一定灵活性的。与编辑的信件来往非常重要，可以说不亚于稿件本身。编辑不一定会阅读稿件的全部，但会阅读信件的全部。信件的水平能反映出稿件的水平和作者的可信度，所以信件的构思、逻辑、语言上都不能马虎。信件的内容要合理、合适，让数据说话，不要带有感情的因素。

稿件被接受后，最后一步是出版程序，如校对、签版权合同、购买单行本等。校对的最好办法是一个人念稿件，另一个人核对，这样最容易发现问题。这个阶段的时间要求比较紧，作者应及时、认真地完成期刊社的要求，使论文顺利发表。

2.1.1 稿件评价标准

目前，同行评议 Peer review 已成为学术出版中一项标准流程。同行评议一般为自由评议，一般按着稿件的章节，比如，摘要、前言、讨论、图表等顺序进行评审。但有些期刊会给审稿人提供一些审稿标准，请审稿人逐一评价。这些都是一些基本的要求，作者可以通过对照这些审稿标准，检查自己的稿件能否过关。

下面列出了一些通常的要求。

（1）论文研究和结论是否对某个技术或科学方面做出了新的进展。Do the research and the results present a significant new advance to a clearly identified technological problem or scientific question? 新颖性、新发现和亮点是一篇论文的核心。

（2）论文的研究目标是否清晰。实验是否与论文目的一致。论文目的应该是贯穿整篇论文的主线。Are the objectives of the research clearly defined, and is the research in line with the objectives?

（3）论文的标题是否准确反映了论文的主要研究。Does the title adequately reflect the content of the manuscript?

（4）摘要部分是否能单独概括整篇论文。即使不通读论文也能知道科研的目的、方法、结论和意义。Can the abstract be understood on its own and does it contain the relevant facts?

（5）前言部分是否围绕研究目标而清晰表述，是否包括科研背景、目的、意义等。Is the introduction clearly presented, and confined to the research objective?

（6）实验方法是否合理，步骤是否齐全，是否定量，能否达到被重复的要求。Are the data quantitative, understandable, and presented clearly? Is the statistical treatment of the data adequate? Are the experimental methods appropriate, and are they described clearly and with sufficient details for the work to be repeatable?

（7）结果部分是否清楚表述，数据是否支持结论。Are the results and conclusions clearly presented and do they support the claims made by the authors?

（8）讨论部分是否围绕论文的新发现、贡献、意义、缺陷和应用前景，或者潜在应用展开。Is the discussion integrated with the results, clearly presented and confined to the research data?

（9）图和表格是否清晰、质量高，是否需要增加或减少图和表格。Are the figures and tables of high quality?

（10）主要文献是否都被引用，是否引用过多文献，文献写法是否准确。Are the most relevant references all cited, can unnecessary or marginally relevant references be omitted?

（11）英文写作、语法、用词，是否达到标准，是否需要专家润色。Are the language, grammar, and style of high standard?

（12）稿件格式是否符合期刊要求。Are the formats in line with the requirements of the journal?

2.1.2 Cover letter 的写法

投稿时一般要附上一封给期刊编辑的 Cover letter，信中简要描述论文的亮点和意义，为什么要投给这个期刊，做到简明扼要。如有什么要求，可以在信中提出，如不希望某些同行看到原稿等。有些期刊要求作者建议审稿人，这也需要在信中列出。Cover letter 一般不必写太长，但要把稿件的精髓和意义写出来，避免照抄论文摘要，语言要准确，专业。一个简单的 Cover letter 一般可以这么写：

例 1.

Dear Editor:

We would like to submit our manuscript entitled "Convenient and Diastereoselective Diacetoxylation of Alkenes Mediated by $PhI(OAc)_2/BF_3 \cdot OEt_2$ System" (标题) for publication on *Journal of Organic Chemistry* (期刊).

We describe a "metal-free" and highly efficient method for the diacetoxylation of alkenes mediated by $PhI(OAc)_2/BF_3 \cdot OEt_2$ system. By using this methodology, not only syn-diacetate products but also trans-diacetate products were achieved in good to excellent yields with high diastereoselectivity simply by carrying out the reaction in the presence or in the absence of water. A broad range of substrates are compatible with this novel method, and even electron-deficient alkenes (such as $α, β$-unsaturated esters) are dioxygenated efficiently to produce the desired products at room temperature. Moreover, multigram-scale reaction is also successfully carried out with comparable yield and diastereoselectivity to small-scale reaction, demonstrating the high efficiency of this method. We expect this mild, practical and novel procedure would be useful for the preparation of various 1,2-diols in organic synthesis (论文的创新点和意义).

The work described was original research that has not been published previously, and not under consideration for publication elsewhere, in whole or in part. We greatly appreciate your time to review our manuscript and we are looking forward to hearing from you soon.

Sincerely yours,

××××

例 2.

We would like to submit our manuscript entitled "Expression of an alkaline feruloyl esterase from thermophilic *C. thermophilum* and its enhancing effect on the delignification of pulp" (标题), which we wish to be considered for publication in *Enzyme and Microbial Technology* (期刊).

Exploration of feruloyl esterase (FAE) with the resistance to heat and alkali conditions in biobleaching process to improve the separation efficiency of lignocellulose is the key to achieving green papermaking. In this report, we expressed FAE of *C. thermophilum* and obtained a thermostable alkaline FAE that can effectively promote the removal of lignin from pulp. The recombinant CtFAEB has broad pH stability and can retain about 56% of the maximum activity even at pH 11.0. Alignment of the protein sequences of CtFAEB and mesophilic FAE suggested that the percentage of amino acids that easily form alpha helix in CtFAEB increases, which enhances its structural rigidity and thereby improves its thermal stability and alkali tolerance. Our study provides an effective method to obtain thermostable and alkaline FAEs, which will promote its application in biorefining industries (论文的创新点和意义).

Due to these important observations, we are confident that our manuscript meets the acceptance criteria of *Enzyme and Microbial Technology*. We are looking forward to hearing from you soon.

Sincerely yours,

××××

2.2 稿件重修

论文投稿后的最好结果就是可以发表，但需要重修。重修也可能是小修，也可能是大修。不管怎样，修改后的稿件大概率会被期刊接受发表。当文章回来审稿意见后，作者应该按照审稿人意见进行修改，尽量不再作多余的改动。在稿件重修过程中，对于审稿人要求修改的内容，一定要认真修改，不能敷衍，争取一次重修就能过关。重修的稿件，期刊一般要求把修改处标记清楚，以便于编辑和审稿人查看是否已按要求进行了修改。如果一些审稿意见没有进行回复，作者很可能就会收到这么一条意见："The authors did not consider the suggestions for improving the manuscript. Besides, they did not answer some questions asked"。如果这样的话，期刊编辑能让你的稿件过关吗？

稿件重修时需要认真分析审稿意见，审稿人意见一般分为这么几类：

（1）要求补充实验和数据。即现有数据不能很好地支持论文的结论。

（2）要求改进章节中的表述。审稿人对于某个章节（sections），比如 Abstract、Introduction、Methods、Discussion 等提出意见，要求改进写作内容。比如，增加或删掉摘要的某些内容，提高前言的逻辑性和突出研究意义，增加对论文结果和已发表的工作的对比，等。

（3）要求改善实验和数据的表述。对某个实验、数据或结果有疑问，或认为表达不清楚，要求改进。

（4）要求改正不准确的语言问题，比如语法和难懂的句子等。

（5）要求降低稿件重复率。如果查重率太高，期刊编辑会要求修改以降低重复率。

这几类问题中，要求补数据是最重的，但也必须照做。如果已经不可能补数据，比如样品没有了，应该看看能否找到文献数据支持。同时，给编辑答复时，一定要把情况说明，再加上文献支持，也许能得到审稿人或期刊的认可。但大部分情况都需要进一步补充真实数据。其他几类意见相对来说比较容易处理，但也要认真按照审稿人意见进行稿件修改。

在递交重修后的稿件时，需要写一份答复信 (response letter)。信中要把所有改动都一一列出（point-by-point）。所以答复信就类似一个小稿件，要同写稿件一样认真对待。一般来说，对稿件的修改，在答复信中都能反映出来。也就是说，写好答复信和修好稿件二者是统一的。下面针对不同类的审稿意见，结合范例对答复信的写法进行详细介绍。

2.3 答复信的写法

答复信是写给期刊编辑的，要有一个简洁的开头。一般先感谢编辑的工作，但也不要过于客气。对编辑和审稿人的要求和评语要一一答复。然而，一些作者在写答复信时，经常会存在一些问题，这些问题虽然不太会影响整个答复信，但可能会在交流或者语气上，对期刊编辑造成一些小的负面影响。撰写答复信时的常见问题如下：

（1）没有针对意见答复。审稿意见大多是英文，由于语言问题，常常会造成理解错误，或者，也可能是作者想敷衍过去。造成的结果就是回复的审稿意见明显答非所问。期刊编辑的工作就是把关好每一篇稿件。如果作者无法针对审稿意见给出合适的答复，是无法在期刊编辑这里被放行的。

（2）格式不是很合理。答复信虽然没有像稿件那样有严格的格式要求，但一般都是写给编辑，表示感谢，说明稿件已按意见修改，并将修改处在此信后面一一列出，希望能被期刊接受，以及作者的署名。有的作者常常会把给编辑的开头信和署名放在后面，这样的格式就不太符合国际上的通常做法。

（3）过分感谢或过分谦虚。比如，每一个答复，即使是一个很小的建议都以"Thank you very much"开始。一般来说，对审稿人表示感谢一次就可以，没必要句句感谢。同时对自己的粗心错误，表示一次歉意就可以，也没必要句句表示道歉。

（4）答复信的英文要尽量标准，准确。答复信的质量反映稿件的写作水平，好的答复信有助于论文被接受。

下面是几个答复信的范例：

例1.

Dear editor:

We appreciate your positive and constructive suggestions on our manuscript. All revisions are color marked in the revised manuscript. The point-by-point responses and revision details are listed below. Thank you very much for your time and efforts on our manuscript and your kind consideration!

Sincerely,

××××

例2.

Dear editor:

Thank you and the reviewers for your valuable suggestions. We have carefully read through the comments and made proper revisions. In addition, the writing was edited by an English editor. Our responses to the reviewer's questions are listed below. We greatly appreciate your time and efforts to improve our manuscript for publication.

Sincerely,

××××

例3.

Dear editor:

We are thankful for the positive and valuable suggestions for our manuscript. As the reviewers suggested, we have improved the language and corrected the typographic errors. Also, the issues regarding the format have been resolved. The point-by-point responses to the comments are listed below. Thank you again for your time and efforts and we hope the revised manuscript can be accepted for publication.

Sincerely,

××××

例 4.

Dear Editor:

We sincerely appreciate the constructive comments from you and reviewers, which are valuable for improving the quality of our manuscript. We have revised the manuscript according to the reviewers' comments. We hope, with these modifications and improvements based on your suggestions and the reviewers' comments, the quality of our manuscript would meet the publication standard of clinical cardiology. The responses are listed as follows. Changes are highlighted in the manuscript.

We are looking forward to hearing from you.

Sincerely,

××××

例 5.

Dear Editor:

Thank you for the opportunity to publish our revised manuscript "××××" in *Medicinal Chemistry*. We appreciate reviewers' time and efforts in reviewing our manuscript and are grateful for the insightful comments and constructive suggestions, which significantly improve the quality of our manuscript. After carefully reconsidered the comments of reviewers and revised our manuscript in details, we summarized our responses to reviewers' comments as follows and marked the changes in red in the manuscript.

We would like to thank the reviewers and editor for taking the time to review our manuscript again.

Sincerely yours,

××××

2.4　一些常见问题的答复范例

2.4.1　对要求补充实验或数据的答复

例1.

"I am afraid that a clear proof that the agents are binding specifically to Mcl-1 is still missing. Adding the agent to the protein causes the CD spectrum to look less helical. Is the agent denaturing the protein or aggregating it? Also, adding more agent caused more changes in the CD also past 1:1 at 2 μM, which is unexpected for a low nanomolar agent."

我担心论文仍缺少证明化合物与 Mcl-1 蛋白特异结合的明确证据。试剂与蛋白质的结合造成 CD 光谱中螺旋结构变少，是否是因为试剂使蛋白质变性或聚集造成的？

Response: Thank you for the comments. In the revised manuscript, the SPR direct binding assay was re-done. We optimized the experimental method. The new result was shown in the newly created Table 2 and Figure S47 and discussed in Section 2.3. The new data fit well with the 1:1 binding model. In addition, we corrected the wrong description in the text. In the revised manuscript, the cellular thermal shift assay (CETSA) was adopted. CETSA is a direct binding assay that monitors ligand-mediated thermal stabilization of target protein in cells. The result was shown in Figure 3 and Figure S50 and discussed in Section 2.3. Complex 14 significantly enhanced the thermal stability of cellular Mcl-1 protein at low nanomolar concentrations (EC_{50}=6.3nM), which is expected for a low nanomolar agent.

例2.

"Are the compounds fluorescent? By the way, the supporting information doesn't contain any information about this. These signals can give a lot of information about what is happening in solution."

所有化合物都有荧光特性吗？此外，支撑材料没有包含这方面的任何信息。这些信号可以提供大量的关于溶液中发生情况的信息。

Response: We tested the fluorescence of all the complexes in solution and the data were added in supporting information Figure S2. The data indicated that fluorescence peaks of complex 1 were obviously different from other complexes in both DMSO solution and DMSO-PBS solution, likely affected by the counter ions (sodium) in PBS. These results suggested that the fluorescence of the complexes 1–6 was mainly affected by the ligand H—L.

例 3.

"It is better to add real data of demand and utilization of CBM in other areas to verify the conclusion in this paper, so as to increase the reliability of conclusion in this paper."

最好能加上其他领域对 CBM 需求和利用的现场真实数据，来验证本论文的结论，从而增强本论文结论的可信度。

Response: Thank you very much for your comment. It is indeed worth discussing the applicability of the findings of this study in other mining areas. But the main purpose of this research is to study the characteristics of coalbed methane resource demand and resource allocation patterns in mining areas, and to solve the problem of resource allocation in mining areas. Although the research results are based on the data from YM mining area, the research problems exist in most mining areas, and there are certain differences in the coalbed methane extraction capacity and utilization policy of each mining area. Therefore, the research methods and results used in this article can provide references for other mining areas and cannot be copied completely. On the other hand, it is difficult to collect a large amount of data required for CBM research in other mining areas. Moreover, there are relatively few studies on the supply and demand relationship and resource allocation of CBM, and there is a lack of data that can be cited.

例 4.

"The PK studies showed very small amount of circulating agent (double digit nanomolar) yet the agent has some efficacy in *vivo*. The authors responded to rev 2 that the compound accumulated more in the tumor with new data, yet the concentration in the tumor is still nanomolar, while cellular activity required to see efficacy $100\times$ $-1000\times$ these concentrations. I am also still puzzled that T_{max} is greater than the half-time (how is that possible?)."

药动学实验发现循环药物浓度很低（两位数的纳摩尔），但该药剂在体内却有一些疗效。作者在第二次修改中用新的数据回复了审稿人，表明化合物在癌组织中会积累更高的浓度，但药物在癌组织中的浓度仍然很低，而达到细胞活性的药效浓度是这些浓度的 1007–1000 倍。我还对 T_{max} 大于半衰期不能理解（怎么可能？）。

Response: We agree with reviewer's suggestions. 1. The pharmacokinetic studies were re-done. We optimized the detection method and the solvent. The new data were shown in Table 6 and Figure S52 and discussed in Section 2.7. This drug concentration is much higher than the cellular IC_{50} values, indicating complex **14** has an adequate exposure in mouse model. 2. The tissue distribution analysis was re-done. We optimized the detection method and the solvent. The new data were shown in Figure 10D and discussed in Section 2.9. The drug concentration of complex **14** in tumor tissue was higher than the cellular IC_{50} values, indicating that complex **14** has an adequate exposure in tumor tissues.

例 5.

"The lethal dose of the agents is reported to be about 30 mg/kg, which seems way too toxic for an agent that only works on micromolar level in cell."

论文报道的致死剂量约为 30 mg/kg，这对于细胞中需要达到微摩尔水平才具有活性的药剂来说不合理。

Response: We agree with reviewer's suggestion. This experiment was re-done. We optimized the solvent. The final concentration of DMSO was limited to 1%. The new result was shown in Figure 8 and Table S13. The new LD_{50} values of our copper complexes were much higher than the dose required for cancer treatment in xenograft mouse model.

例 6.

"Add the analytical method validation data (also as supplementary), reporting also the chromatograms in order to check the selectivity and sensitivity"

补加分析方法验证数据（也可作为支撑材料），也需要补加关于选择性和灵敏性的图谱。

Response: This paper mainly describes the study on the bioequivalence of two injections, among which the methodological study on the determination of drug concentration is not the focus. We have provided data related to methodology validation studies in this response letter (see Table. Methodology validation data), but we do not think it is necessary to add these results to the manuscript.

例 7.

"All of the intermediates and probe HBTMP should be characterized by ^{1}H NMR, ^{13}C NMR, FT-IR and high resolution NMR spectra should be provided."

所有中间体和产物都应通过 ^{1}H NMR、^{13}C NMR、FT-IR 进行表征，也需要补充高分辨率 NMR 图谱。

Response: Thank you very much for your suggestions. The high-resolution NMR spectra and FT-IR of HBT-CHO and HBTMP have been replaced. Please see Fig.S8-10.

例 8.

"The reaction substrate is not enough. Authors can try more large steric hindrance such as *tert*-butyl or other substrates."

反应物的种类不够多，作者应试试有更大空间位阻的基团，例如叔丁基或其他反应物。

Response: As suggested by the reviewer, the reactions of 4-(tert-butyl)-o-phenylenediamine and 2,3-diaminopyridine with CO_2 in the presence of BH_3NH_3 were carried out under the optimum conditions. The product 5-(tert-butyl)-1H-benzoimidazole (**2f**) was obtained in 84% yield; however, no target product was detected when 2,3-diaminopyridine was used as a substrate, presumably because the coordination of nitrogen atoms in 2,3-diaminopyridine with BH_3NH_3 decreased the reactivity of the nitrogen atoms of amino group.

例 9.

"Their improved procedure somewhat increased the yield of this ligand, but scientifically resembles to the previous work."

论文改进的方法只在一定程度上提高了这种配体的产量，但科学方法上与文献报道没有太大的区别。

Response: We disagree with your assessment of the presented work. Compared with the previous literature, our work has made significant improvements. In the introduction part of the manuscript, we elaborated the problems existed in the literature, as described below.

…(省略)

In view of these limitations, we analyzed the causes of the problems and put forward feasible solutions. We performed experiments to address these limitations and these improvements have never been reported in the literature. For example:

…(省略)

We believe our study provides many useful and new data for the laboratory use and industrial application of chiral water-soluble sulfonated BINAP derivatives in asymmetric hydrogenation.

例 10.

"This work would have also benefited from kinetic and substrate specificity studies."

提供动力学和底物特异性的研究。

Response: Four model substrates have been used to determine the kinetic constants and substrate specificity. The experiment methods and results have been supplemented in the section 2.6.3 and 3.3 of the revised manuscript.

例 11.

"It would be better to provide the selected electron diffraction results from TEM test to further confirm the amorphous and no crystals state of species on electrode."

提供 TEM 实验得到的电子衍射数据，用以验证电极材料的无定形状态。

Response: This compound has been well studied as an amorphous electrocatalyst. This material has been used in photoelectrochemistry for many years. Their amorphous structures are demonstrated mostly by XRD and SEM in the literatures …(省略). This compound is the amorphous crystal directly deposited on a carbon matrix in our work, which is incompatible to TEM imaging. Therefore, its amorphous and non-crystal state is usually not characterized by TEM in literature. In our work, we demonstrated the amorphous state of this catalyst by XRD and SEM based on the properties of the electrode and literatures.

例 12.

"It is not convictive to conclude that increasing hydrophilicity of electrode surface can promote interaction between membrane protein and the electrode surface, because there is not any data about membrane protein in results and discussion section of this manuscript."

提高电极表面的亲水性有助于增强膜蛋白和电极的相互作用的结论没有说服力，因为稿件的结果和讨论部分没有任何关于膜蛋白的数据。

Response: We have conducted additional experiments for the interactions between membrane protein and the electrode surface using the bacteria with and without membrane protein in the revised manuscript to demonstrate that increasing hydrophilicity of electrode surface can promote its interaction with membrane protein. The experiment methods and results are present in the supporting information.

2.4.2 对要求改进章节中表述的答复

例1.

"An abstract should be inclusive of all the evidence and data that support the interpretations and statements. For example, using of "aside from other petrographic observations" is not informative. One should understand the abstract without going to the text for further information."

摘要部分需要改进,要列出主要数据和证据,用以支持论文的结论。例如,"除了其他岩石学观测之外"这个句子中没有什么有用的信息。要让读者不用看全文就能理解摘要。

Response: The abstract is now completely rewritten and we appreciate the comments from the reviewer.

例2.

"The references need to be carefully checked. For examples, page numbers are missing for Bachmann et al. in Line 773, Bowen 1928 in Line 785, and in Line 820. Volume and page numbers are missing in Line 980. This indicates that more work is needed to improve the manuscript, including all the other sections."

文献部分需要仔细检查,多个地方缺少页码。这也表明整篇稿件都需要改进。

Response: Corrections have been made as suggested.

例3.

"The abstract of the article is too long, so it is suggested to simplify."

文章摘要部分太长,请缩短。

Response: Thank you very much for your suggestion. We have condensed the abstract and highlighted the background, methods, results and significance of the research. For details, please refer to the revised summary. The relevant changes are marked in red font.

例4.

"The authors presented results during introduction, it is not recommended."

作者在前言部分描述了结果,不符合论文格式,应予改正。

Response: We thank reviewer's suggestion. The results are only presented in Result section in revised manuscript.

例 5.

"Abbreviations should be defined in the text when the first time used and their use should be minimized. The symbols used in equations should be defined following the equation and they should be provided in the nomenclature together with respective SI units."

缩写在第一次出现时要有全称，并且也应尽量避免使用，方程式的符号要在方程后说明，并应在术语中加注这些符号和对应的 SI 单位。

Response: Thank you very much for your comment. We have reduced the use of abbreviations in the manuscript and standardized the format of the equations. The corresponding changes are marked in red.

例 6.

"The Abstract should be concise and inform the readers of background, research question, hypothesis, methodology, the main results and conclusions of research presented, and ideally, the main implications and broader context of your findings."

摘要要简洁描述科研的背景、解决的问题、假说、方法、主要发现和结论，最好还能包括研究结果的主要影响和更广的背景。

Response: We have reorganized the abstract and clarified the background, methods and conclusions of the research. For details, please refer to the revised abstract.

例 7.

"The Introduction should be written starting from a broader context of your research, leading readers to more specific aspects, and the research gaps which you are addressing should be clearly identified."

前言部分应从大面（的背景）开始，延伸到论文的聚焦点和论文要解决的具体问题。

Response: We have revised the introduction and improved the research background. The corresponding changes in the introduction are marked in red.

例 8.

"The novelty should be clear and additionally highlighted, together with the objectives of your research, in the last paragraph of the introduction. A section explaining organization of the paper is a redundancy and should be avoided."

在前言的最后一段，应讲述论文发现的新颖之处，同论文研究的目的相对应。

解释论文结构的部分是多余的，应该去掉。

Response: The last paragraph of the introduction has been revised in accordance with your comments, highlighting the innovation and purpose of the research.

例 9.

"Your previous relevant work and work from other researches should be properly cited and relevant references used. Findings presented elsewhere and used in your work should be summarized and written in your own words, and the original sources should be cited."

你已发表的工作和他人的工作应适当地引用并加注文献。要用自己的语句来概括和描述你工作里用到的和其他地方提到的发现，并附加来源。

Response: Thank you very much for your comment. We have cited the previous related research of our research group and improved the description of the current research status in related fields. The corresponding changes in the introduction are marked in red.

例 10.

"Lumping references (citing several papers together) should be avoided and such sections should be rewritten to be clear what the main contribution of each cited paper is."

应避免一起引用多篇文献，引用多篇文献的句子要重写，清楚表明每篇文献的结论（主要贡献）。

Response: Thank you very much for your comment. We have made reasonable changes to the citation of references to avoid quoting multiple citations at the same time.

例 11.

"The methodology should be concise and logical allowing interested researchers to be able to repeat your work. If the methodology, or some parts, have been already published elsewhere, you should summarize it and provide reference."

方法要简洁和有逻辑性，以让感兴趣的读者能够重复。应简述已经发表的方法或其中一部分，并提供文献。

Response: Thank you very much for your comment. I have simplified the research method, reorganized the research method, and deleted unnecessary content. Corresponding references have been added for readers to further understand, and the corresponding changes in chapter 2.1-2.3 are marked in red.

例 12.

"The results should be quantitative, discussed and compared with the results published in the literature, and speculations should be avoided. It must be clear what results are original and presented for the first time and what is used from the literature for comparison, providing references."

要有定量的结果，并与发表的文献进行比较和讨论，同时应避免猜测。要明确表明哪是新发现，哪是文献报道，并提供文献。

Response: Thank you for your comment. We have revised the results and discussion of this article as required, and removed speculative language to ensure that the conclusions of the article are based on data.

例 13.

"In addition to the main findings, the conclusions section should indicate research gaps and research directions identified as the results of research presented."

除主要发现外，结论部分请补加本工作的不足和将来的研究方向。

Response: Thank you very much for your comment. We have pointed out the research gaps and future research prospects in conclusion (3) based on the current research results.

例 14.

"The significance of the work is not made clear. This appears to replicate a subset of the findings of Zhou et al (23). The contribution of this work is not obvious."

这项工作的意义尚不清楚。像是复制了文献 (23) 的部分结果。这项工作的贡献并不明显。

Response: In the 3D-ROI segmentation method by Zhou et al, the tumor segmentation was done on each 2D slice, and they were rendered into a 3D space with isotropic voxel resolution for extracting the 3D texture features. In our study, a physician with professional breast MR diagnosis experience performed manual segmentation. Firstly, they segmented the 3D-ROI based on the tumor itself slice by slice along the edge. Then, the maximum diameter of the lesion is projected onto 3 coordinate axes of the image to determine its coverage range of x, y and z axes. The bounding box of the tumor is finally obtained.

例 15.

"Please move the Experimental Section of the manuscript to appear after the Introduction and before the Results and Discussion section." "Results section is very poor, need to rewrite in a good way and numbers."

请把实验部分挪到前言之后，结果和讨论之前。结果部分很乱，需要重写，用小段落分层。

Response: The experimental section has moved to the suggested place, after the Introduction and before the Results and Discussion section. We also made major revisions in the Results section.

例 16.

"It is not persuasive to conclude that rougher surface is conducive to electron transfer between bacteria and the electrode as mentioned by the author. Please provide the solid evidence or relative reference."

电极表面的粗糙性有利于细菌和电极之间的电流流动这一结论不具说服性，请提供坚实的数据或文献支持。

Response: We may not express it very clearly, so we have rewritten it and added relative reference in the revised manuscript.

例 17.

"The authors cited too many references in the Introduction. It's not a comprehensive review. You can just show the most related ones."

作者在前言中引用太多文献，这不是一篇综述，应该只引用最有关的文献。

Response: We have double checked all references, and removed the references not closely related to our work.

2.4.3 对要求改善实验或数据的答复

例 1.

"The figure quality needs to be improved and error bars should be added in Figures."

图表质量需要改进，应该在图中补加误差线。

Response: The figure quality has been improved and we added error bars in Figures. Please see Fig.2.

例 2.

"Fig. 1. please add scale to WWZ image to be able to appreciate distances. The white square on the large map seems larger than the inset of HT Bay, please adjust. Please provide source of the maps (Google Earth)."

加图像尺度，调整相对比例，提供图像来源（谷歌地球）。

Response: As suggested, distance scale and map source were added and other revisions were also made in Fig. 1.

例 3.

"Figure 3: It is unclear how the flocculation efficiency was calculated. Please include more details in the methods section."

不清楚絮凝率是如何计算的，请在方法部分表明。

Response: Thank you for your suggestion. The flocculation efficiency was calculated as described previously: Flocculation efficiency %= $(A-B)/A*100$，where A and B are the absorbance at 750 nm of the bacteria culture in control and sample, respectively. We have added more details in the methods section.

例 4.

"Figure 2- Figure 5 and Figure 7- Figure 8 are not standard. It is recommended to modify them."

图 2-图 5 和图 7-图 8 制作不标准，建议修改。

Response: Thank you very much for your careful inspection. We have modified the format of Figure 2, Figure 3, Figure 4, and Figure 7. However, since Figure 5 and Figure 8 are the residual autocorrelation and partial autocorrelation graphs, which are obtained from the SPSS prediction software, the format of the plotted graphs cannot be changed. If the data is exported and graphed, the range of the confidence interval is difficult to determine. We have also consulted many published papers, and many scholars have directly used graphic files in this format.

例 5.

"On the basis of the real data, the demand and utilization of high and low concentration of CBM in peak and low peak periods should be analyzed."

根据现有数据，补加对高峰和低峰阶段 CBM 高低浓度的需求和利用的分析。

Response: Thank you very much for your suggestion. We have analyzed the characteristics of changes in demand for high-grade coalbed methane and low-grade coalbed methane in Section 3.1, and the corresponding changes are marked in red.

例 6.

"It is suggested that Table1 should be illustrated by figure."

把表 1 改为图。

Response: We have merged the contents of Table 1 into Figure 3 and Figure 4. In Section 3.3, the corresponding table information has been merged into Figure 7.

例 7.

"The substrates used in Scheme 2 are ambiguous, which group does R represent?"

方案 2 中的反应物不清楚，R 代表哪个基团？

Response: We have rearranged Scheme 2, in which the substituted group of the product is corresponding with the starting materials.

例 8.

"The C–F bond coupling constant for product 3f is incorrect."

产品 3f 的 C–F 键的耦合常数错误。

Response: Thank you for your correcting the mistake of our manuscript. After the NMR data was checked carefully, it was found that the C-F bond coupling constant is 239.65 Hz. (See 3.7 in ESI, 5-fluoro-1H-benzoimidazole **2g**)

例 9.

"As regards the antiproliferative activity, the authors should also present data for noncancerous cell lines. Any compound cannot be considered a potential antitumor drug unless it is preferentially active in cancer tissues."

作者需要提供对非癌症细胞的毒性，任何化合物都不能被认为是一种潜在的抗肿瘤药物，除非它在癌组织中具有优先活性。

Response: The IC_{50} values of noncancerous cell lines (HL-7702) were presented and discussed in detail in the manuscript (page 12-13).

例 10.

"The authors must assign all the protons and carbons possible using 2D techniques, for that the atoms of the ligand must be labeled in scheme 3."

作者需要用 2D 技术标注每个氢和碳原子的峰，必须在方案 3 中标记出配体的每个原子。

Response: We have assigned all the protons and carbons possible using 2D techniques (HSQC and HMBC spectra) as labeled in Scheme 3, which are shown in detail in the Electronic Supplementary Information (page S14-16).

2.4.4 对要求改正不准确语言的答复

例 1.

"The English language should be at required standards, without typos and grammar errors. Too long sentences difficult to follow should be avoided. Ideally, it should be edited by a native English speaker."

要用标准英语，不能有拼写错误和语法错误，避免难懂的长句子，最好有母语编辑润色。

Response: Thank you very much for your comment. We have contacted a professional editing service to carefully check the language of the article to ensure the quality of the article.

例 2.

"Language needs major improvement."

语言需要大幅改进。

Response: We apologize for the language problems in the manuscript. We have asked a native English speaker to go through the article.

例 3.

"Past tense has been used extensively for verbs in the text and in many cases it should be just verbs. For example, in line 353, "they displayed enrichment in light REE" should be "they (the porphyry samples) display...". These are the rocks' current features. This issue should be corrected throughout the text."

过去式和现在时用法混乱，给出了举例。整篇需要改正。

Response: This might be a confusion that needs to be explained. We are following the standard style guide that all facts are presented in present tense and all results and findings of this study are given in past tense. Hence, if we ran an experiment to find out a particular characteristic of a rock, the verb would still have to be "had" since this property was only newly determined from our experiment. In the new revision, corrections have been made throughout the text.

例 4.

"Line 28-29, Try to reword "them", "they", and "their", so that readers know which plutons are being referred to, as there are four in the study."

代词 "them" "they" and "their" 代表对象不清楚，此研究中有四个研究对象。

Response: Corrections have been made in the revision.

例 5.

"Writing style and grammar are also a bit too deficient and would require significant improvement."

英文写作和语法比较差，需要大幅改进。

Response: We revised this manuscript and corrected the grammar mistakes.

例 6.

"Line 46, delete "that occurred"; Line 63, can be easily melted; Line 64, cannot be partially melted; Line 69, delete outcropped; Line 71, became the → became a; Line 84 delete the first "the", as this is the first appear of the plutons studied."

建议了具体的英文修改意见。

Response: These corrections have been made as suggested.

例 7.

"English should be reviewed (in line 46 "under fasting conditions") or the abbreviations should be specified the first time that appeared in the text."

英文需要润色，缩写第一次出现时要注明全称。

Response: The English has been edited by a native speaker. The abbreviation has been rechecked and confirmed to be specified at the first time appeared in the text.

例 8.

"It is recommended that native English speaking institutions or experts polish this article."

建议寻求母语专家润色。

Response: Thank you very much for your comment. We have sought professional editing service to check the language of the article.

例 9.

"Please define at-fist all used abbreviations (DEG's, FoxO, etc.). Please include the OMIM IDs to all genes and phenotypes mentioned at-first. Also, the review of the manuscript by an English grammar editing process would be desirable."

缩写第一次出现时要注明全称，第一次描述基因时要注明 ID，建议寻求专业编辑服务。

Response: We have checked and modified all abbreviations and added OMIM IDs to all genes and phenotypes. The English grammar and writing are checked by a professional editor.

例 10.

"Please check the English of the manuscript carefully. There are some spelling and grammar errors. For example, 1) Page 2, line 33, "utilization" should be "utilized"; 2) Page 13, line 33, "the 1a" should be "1a"; 3) Page 14, remove "the" from "the CO_2" and "the SH"."

仔细检查稿件的英语，稿件中存在许多拼写和语法错误，并附加具体示例。

Response: Thank you for correcting the spelling and grammar in our manuscript. According to your suggestion, we had checked and corrected the spelling and grammar throughout the manuscript.

2.4.5 查重率太高问题

高查重率一般是照抄文献或重复了自己发表的论文而造成的。为降低查重率，对相同的实验数据或结论可换一种方法描述，如替换同样意义的词组、调整句子结构、调换词组位置等。下面举些例子，斜体部分是与文献重复的。

Exosomes appear as vesicles *with a size range of 30-150 nm*.

宜改为: Exosomes appear as vesicles *with sizes in the range of 30 nm to 150 nm*.

The activity increased *in a concentration-dependent manner*.

宜改为: The activity showed *concentration dependent increase*.

It circulates *in almost all body fluids (e.g., blood, urine, saliva, and breast milk, etc.)*.

宜改为: It circulates in *blood, saliva, breast milk, urine, and almost all other body fluids*.

To improve the cytotoxicity of the Cu complexes we prepared previously, we synthesized the new compounds C1-C5 in the study.

宜改为: *Our previously reported Cu complexes showed moderate cytotoxicity; therefore, new compounds C1−C5 were prepared in this report to improve their activity.*

Recently, as a potential supplement to the nanomaterials-based TCD agent, metal ligand complexes have attracted extensive attention.

宜改为: *Metal and ligand complexes have recently attracted considerable attention as a potential replacement to the TCD agent made of nanomaterials.*

第3章
英文写作

文字是信息的记载和传播形式，要达到信息的准确传播，论文写作的最终要求是读者对文字中传达的信息的理解要与作者希望传达的信息完全一致。语法也就是语言写作的规则，文字要按统一的语法规则来排列才能写出相互理解的文章。有的语法对科技论文写作可能关系不大，如将来式的用法，但同样有些语法对科技论文写作特别重要。

清晰的实验描述和准确的思想表达对科技写作尤其重要。宁愿多写一些句子，也要把事情表达清楚，也就是不能为达到简洁而牺牲清晰。句子可以写得简单，用词也可以有些单调，但语句的意思一定要清楚。科技写作的语法特点就是为了保证语句的清晰，例如"new student group"是表示"new STUDENT GROUP"还是"NEW STUDENT group"，二者还是有很大差别的。一个语句有多种解释或含糊不清是不可以的，例如："as discussed above"，你自己知道 above 指什么，但读者可能不清楚，或需要到前面寻找答案，这就是不清晰，这时需要明确地写出你要指的事情。再例如："under the new condition the yield increased significantly" significantly 是比较含糊的一个字，应该写出产率增加了多少，比如"increased by 50%"。

3.1 论文写作中常出现的语法问题

不同人对英语语法的掌握程度不同，下面介绍一些科技写作中经常遇到的语法问题。

英语要求在句子的构成中，不同的组成单元之间要一致，也就是一致性原则（the rule of agreement）。要保持一致的因素有很多，科技写作中经常会出现问题而特别需要注意的有：主语和谓语的单数和复数要一致；修饰语与主语名词要一致；主语和主语的行动（谓语）在逻辑上要一致；代名词和其代理的先行词要一致。

3.1.1 主语和谓语的单数和复数要一致

英语中名词有它的单数和复数形式，动词也有它的单数和复数形式，二者要一致。单数主语（subject）名词要用动词（verb）的单数（singular）形式，复数主语名词要用动词的复数（plural）形式。我们写中文的不太习惯英语的这种写法，很难做到不假思索地配对，需要特别留心才能不出错误，特别是当主语名词和动词被分开时。试看下面的例句：

A high percentage of peptides that are made of amino acids *are* present in the sample.

a high percentage 才是真正的主语，而不是邻近的 amino acids，所以应该用单数形式。

宜改为：A high percentage of peptides that are made of amino acids *is* present in the sample.

让事情更复杂的是英语名词被分为不同的种类，其中的一类叫集合名词。它既可以当单数用也可以当复数用，集合名词当整体来讲时是单数，每个成员作为个体时用复数。例如：

The number of mice in the experiment *was* increased.

A number of mice *have* died.

All of the samples *were* analyzed.

All of the safety procedure *was* strictly followed.

代词 none 既可以是单数也可以是复数。当 none 后面的词是单数时，用单数动词；当 none 后面的词是复数时，用复数动词。

None of the information *was* useful.

None of the animals *were* starved.

描写数量、质量、体积、时间等的词用单数，但如果是分次添加或减少时用复数，在这个意义上与集合名词类似。

1.5 mL *was* added.

10 g *was* added .

6 hours *was* the required incubation time.

5 g *were* added stepwise.

简写的数量单位，如 mg、mL、s 等，单数和复数的写法是一样的，如 1 mg，5 mg。

一些词如 series、type、portion、class，要用单数形式。

A series of derivatives of penicillin *was* prepared.

A large portion of the reports *is* focused on how to deal with the increased cost.

data、criteria、phenomena、media 是复数形式，它们的单数形式分别是 datum、criterion、phenomenon、medium。

3.1.2　修饰语同主语名词关系上要一致

当用动名词（gerund）、分词短句（participle）、不定式短句（infinitive）作修

饰语时，修饰语中的动词要与主句中的主语名词关系上要一致。科技杂志论文中有这种语法错误的情况较多，严格来讲这只是种语法错误，一般不影响对句子内容的理解，所以很多作者不太注意。编辑和阅稿人有时也没有严格要求改正，比如下面就是 Nature 杂志 2006 年第 439 卷中的一个例子。

Using the enhancer GAL4/UAS expression system, short-term memory traces of aversive and appetitive olfactory conditioning have been assigned to output synapses of subsets of intrinsic neurons of the mushroom bodies.

（1）动名词

After finishing the purification, the activity of the isolated compound was then studied.

we or I 是动名词 finishing 形式上的主语，与主句的主语 activity 不一致。

宜改为：After purification was finished, the activity of the isolated compound was then studied.

或：After finishing the purification, we studied the activity of the isolated compound.

Treated with the new drug, the blood cholesterol levels of participants were lowered by an average of 30%.

宜改为：Treated with the new drug, participants showed an average of 30% decrease in their blood cholesterol levels.

（2）分词短句

The iron concentration was determined using the Fenton reaction method.

the iron concentration 与 using the Fenton reaction method 在关系上不一致。

宜改为：The iron concentration was determined by the Fenton reaction method.

或：We determined the iron concentration using the Fenton reaction method.

When measuring the atmospheric level of carbon dioxide, air samples from a remote place, such as an island, is preferred.

宜改为：When the atmospheric level of carbon dioxide is measured, air samples from a remote place, such as an island, is preferred.

（3）不定式短句

To further investigate the potential role of biking in causing infertility, an expanded population of biking athletes was surveyed.

不定式短语的形式主语是 we or I，与主句主语 population 不一致。

宜改为: To further investigate the potential role of biking in causing infertility, we surveyed an expanded population of biking athletes.

To confirm the diagnosis, blood test was ordered.

宜改为：To confirm the diagnosis, the doctor ordered blood test.

3.1.3 主语和主语的行动（谓语）在逻辑上要一致

由于一些中文和英文的表达方式不同,把中文直接翻译成相应的英文会不妥。一个经常被引用的语句是"price is cheap"，中文口语中可以说价格便宜，但其实价格只能高或低，物品才可以说 cheap or expensive。用中文的表达方式来写英文，会出现主语和主语的行动在逻辑上不一致。在写作时要注意行动的真正主语名词是什么，下面是一些例子。

The highest antibiotic *production* was obtained at 48 h.

主语不是 production 而是 production yield。

宜改为：The highest antibiotic *production yield* was obtained at 48 h.

The *scavenging activity* for hydroxyl radicals was based on Fenton reaction.

主语不是 activity 而是 assay of activity。

宜改为：The *assay of scavenging activity* for hydroxyl radicals was based on Fenton reaction.

The *pharmacological* compounds of ginseng were identified.

药物活性化合物应该是 pharmacologically active compounds。

宜改为：The *pharmacologically active* compounds of ginseng were identified.

The *reaction* was freshly prepared and mixed in the proportion of 1∶1∶1 (*V/V/V*) for A/B/C.

主语不是 reaction 而是 reaction solution。

宜改为: The *reaction solution* was freshly prepared by mixing A, B, C in the ratio of 1∶1∶1 (*V/V/V*).

The NMR spectra were *performed* on a Varian 400 MHz instrument.

宜改为：The NMR spectra were *recorded* on a Varian 400 MHz instrument.

3.1.4　代名词和其代理的先行词要一致

代名词和其代理的先行词要在人称、单数或复数，和性别上一致。一些常见的代词是：he, his（阳性单数）；she, her（阴性单数）；it, its（单数）；they, their, these, those（复数）；that, this（单数）。比如下面的例句中，compounds 和 their 一致，protein 和 it 一致。

Many related compounds were synthesized and *their* antivirus activities were studied.

Growth hormone is a protein. *It* promotes human body growth.

下面的例句中，the 应该用 their 取代。

The potential antioxidant capacity of compound A and compound B could be deduced from *the* protective effects against oxidative stresses.

宜改为：The potential antioxidant capacity of compound A and compound B could be deduced from *their* protective effects against oxidative stresses.

用代名词时除了要保持一致外，还要避免代理不清的情况出现，以免不清楚它们到底指什么而引起误解。

The reports were submitted to the requesting agencies, but *they* were ignored.

they 代表 reports 还是 agencies？不明确，容易产生误解。

宜改为：The requesting agencies ignored the submitted reports.

The crude sample was dissolved in water and extracted with organic solvent. *It* was then evaporated to yield the product.

it 指 organic layer 还是指 water layer？不明确，最好不用 it。

宜改为：The crude sample was dissolved in water and extracted with organic solvent. *The organic layer* was then evaporated to yield the product.

又如：In cerebral and heart ischemia-reperfusion animal models, Acid B protected against *these* injuries.

these injuries 不明确，读者可能还要到文章的前面寻找。

宜改为：In cerebral and heart ischemia-reperfusion animal models, acid B protected against *injuries caused by oxidative stress*.

During meal hormones are released after *which* blood flow increases in the stomach.

which 既可以代表 meal 也可以代表 hormones，容易产生误解。

宜改为：During meal hormones are released. After *their release* stomach blood flow increases.

可以把语言看作是字符按一定规则（语法）的排列和组合。使用正确的字符和语法很重要。同样，字、词组或短句在句子中的位置也很重要。组织得当的语言应该能用尽量少的字符传递尽量多的信息。在不增加字符的情况下，包括英语在内的各种语言利用了字词的位置这个变量来增加信息含量。字词在句子中的不同位置，可能传递完全不同的信息，比如"The experiment was done only once"和"The only experiment was done once"，由于 only 的位置不同，所传达的信息就很不一样。英语论文写作中字词在一个句子中的位置关系应注意的有：位置的强调作用，修饰词和被修饰词要邻近，主语和谓语在句子中的位置要靠近。

3.1.5 位置的强调作用

在英语写作中，若要强调某件事情，就把它放在句子的前面。中文写作中，有关句子的条件、时间等的修饰句都是放在前面，而主句总是放在后面。而英文中既可以把条件或修饰句放在前面，也可以放在后面。放在前面就表示要强调修饰句的条件，比如：

Before the hurricane arrived, most of the people have moved out.

Most of the people have moved out before the hurricane arrived.

在英语中两种位置关系都可以，前者强调在 hurricane 来之前，后者强调 moved out。而在中文中，只有一种说法，反过来说"大多数人都离开了在 hurricane 来之前"就不对了。按中文的位置关系直译成英文，往往会不确切，同样按英文的位置关系直译成中文也是怪怪的。例如，"我要吃冰激凌今天，我没吃好长时间了"，就是英文"I want icecream today. I have not eaten it for a long time."的直接翻译，中文要说"今天我要吃冰激凌，我好长时间没吃了"才对。

科技写作中一般还是把主句先写出来，除非你想强调修饰的是条件。

Through scavenging free radicals, antioxidants play an important role in protecting against complex diseases.

宜改为：Antioxidants play an important role in protecting against complex diseases through scavenging free radicals.

In microbial fermentation, phosphorus is commonly the major growth-limiting nutrient.

宜改为：Phosphorus is commonly the major growth-limiting nutrient in microbial fermentation.

Under the current test regime, the engine required 28.5 mg of diesel fuel per cycle.

宜改为：The engine required 28.5 mg of diesel fuel per cycle under the current test regime.

主动句中事情的执行者（作者）放在前面，有强调事情的执行者（作者）的意思，而不是要研究的事物。被动句强调要研究的事物，这也是科技论文中被动句用得比较多的原因之一。

We studied their effects on cell growth 中强调 we。

Their effects on cell growth were studied 中强调 their effects。

3.1.6　修饰词和被修饰词要邻近

科技写作要求严谨，明确，为了严格定义一个事物，往往要加上限制性的修饰词或短句。比如描写实验用的 mice 时，一般不会只说 mice，而是用类似"NCI-H69 tumor bearing female athymic nude mice"的描述，前面有 5 个修饰词来定义研究用的 mice。这时一般把最窄的定义写在最前面，最广的定义写在后面。修饰语要靠近被修饰的对象，因修饰语和被修饰的词被隔开，而造成意思混乱的情况很多，下面是一些例子。

Inhibition of acid B on xanthine oxidase was also reported.

Inhibition of 后面应紧跟 xanthine oxidase，而不是 acid B，隔开后句子就很难读。

宜改为：Inhibition of xanthine oxidase by acid B was also reported.

The chelating activities for ferrous ion of the acid B were assessed.

The chelating activities 后面应紧跟 acid B，而不是 ferrous ion。

宜改为：The chelating activities of the acid B for ferrous ion were assessed.

Reducing power represents the electron donating capacity, which may serve as a significant indicator of potential antioxidant activity.

用 which 开头的修饰句，是要修饰 reducing power，而不是修饰 electron donating capacity，所以要紧跟在 reducing power 后面。

宜改为：Reducing power, which may serve as a significant indicator of potential antioxidant activity, represents the electron donating capacity.

或：Reducing power represents the electron donating capacity. It may serve as a significant indicator of potential antioxidant activity.

Scavenging lipid free radicals of Vitamin E was recently detected.

Vitamin E 是修饰 scavenging，而不是修饰 lipid free radicals。

宜改为：Lipid free radical scavenging by Vitamin E was recently detected.

Administration of B decreased urinary creatol level derived from creatinine oxidation in Wistar rats.

derived from creatinine oxidation 应直接跟在被修饰的 creatol 后面，而不是跟在 level 后面。

宜改为：Administration of B to Wistar rats decreased the levels of urinary creatol that was derived from creatinine oxidation.

3.1.7 主语和谓语在句子中的位置要靠近

要使句子的可读性强，有两个因素需要特别注意。一是句子的长短要合适。研究表明，一个句子中有 13～20 个字时最合适阅读，太短的句子有零碎的感觉，而太长的句子读起来有困难；二是主语和谓语动词要靠近，如果被隔开太远，就会有被隔离的感觉，句子读起来就会比较困难，虽然从语法上来讲是可行的。这主要与人类大脑处理文字信息的过程和方式有关，当人们读到主语时，自然而然地期望知道主语后面的行动，也就是结果。在行动（谓语）出现之前，读者既需要记住主语是什么，同时又要阅读和理解下面的文字，就像要屏住呼吸等待要发生的事情，只有当谓语出现，知道了主题的行动后才能呼出这口气，屏气时间长了自然不舒服。

Lincomycin, one of the lincosamide antimicrobial agents that were first isolated more than fifty years ago, is used as a major antibiotic for the treatment of diseases caused by most Gram-positive bacteria.

宜改为：Lincomycin is one of the lincosamide antimicrobial agents that were first isolated more than fifty years ago. It is used as a major antibiotic for the treatment of diseases caused by most Gram-positive bacteria.

3.1.8　名词作形容词

科技写作中经常会把名词作为形容词使用，如 room temperature，university researchers。当用一个名词来修饰另一个名词时，意义一般都很清楚，但当 3 个名词放在一起，或两个名词前再加一个形容词时，就要小心。有的情况下，3 个或 3 个以上的名词放在一起，表达的意思很清楚，也是一种很简洁的表达方式，如：blood white cell number，prostate cancer patient。但有时会有多种讲法，top university researchers 可以是 researchers of（only）top university，也可以是（all）university researchers who are top。多个名词排在一起，即使表达明确，也给人拥挤的感觉。应避免使用多个名词的修饰方式，最好的办法是用介词或其他方式来把它们分开，以便清楚表达它们的修饰关系。多个名词罗列的情况经常发生，下面多举一些例子。

He wrote the quality control group reports.

宜改为：He wrote the reports of the quality control group.

The patient showed chronic liver disease symptoms.

宜改为：The patient showed symptoms of chronic liver disease.

The human brain oxygen level is quite high.

宜改为：The oxygen level in human brain is quite high.

Their specific inhibition producing effects on fat containing food intake were assessed.

宜改为：Their specific effects of inhibition on the intake of fat containing food were assessed.

The present investigation evaluated various specific drug sample combinations.

宜改为：The present investigation evaluated various combinations of specific drug samples.

The positive drug test result patients were identified.

宜改为：The patients with positive drug test were identified.

Metal chelating property compounds were synthesized

宜改为：Compounds with metal chelating property were synthesized.

Real time PCR selected gene detection method was used.

宜改为：Real time PCR method was used to detect selected genes.

一般做形容词的名词要用单数，除非这个名词总是以复数的形式使用，如 data analysis，sports injury。

cells membrane，宜改为：cell membrane。

nanoparticles suspensions solution，宜改为：nanoparticle suspension solution。

reactions mechanisms，宜改为：reaction mechanisms。

3.1.9 句子的时态

科技论文中基本上只用现在时和过去时两种时态，有的作者偶尔会使用完成时。完成时一般只用于多次并一直在研究的情况，其他的时态用得很少。论文中的时态有它特定的意义，时态用来表明科研成果的认知程度。

当描述已发表的文献成果时用现在时，因为已发表的成果被承认是事实。描述未发表的实验和结果时用过去时，因为还没有得到承认，并且是写论文以前做的事情。由于科技论文中的时态的特定用法，写作中经常需要转换时态。一句话中都可能用两个时态。总起来说，Abstract 中要讲述自己的实验和结论要用过去时。Introduction 中要总结文献和问题，以现在时为主，也用一些过去时。Methods 和 Results 都是讲自己的实验和结果，用过去时。而 Discussion 中则需要根据描述文献还是自己的实验，需要交换使用现在时和过去时。

在引述文献结果时，过去发表的结论可以认为是已经被承认的事实，应该用现在时。但引述过去的实验时，特别是以作者为主语时，因是过去做的事情，应该用过去时。

Wang *showed* that the rate of growth *is* dependent on temperature.

Smith *studied* the growth rate and *reported* that it *is* dependent on temperature.

若作者不是主语而作者的研究是主语时，用现在时。

Investigation by Wang *shows* that the rate of growth *is* dependent on temperature.

当描述自己的实验和实验结果时，应用过去时。因为是在写文章以前做的事情，并且还没被接受为事实（发表）。

We *measured* its plasma concentration and *found* that it *was* two times higher in obesity patients than in normal population.

讲述 table 或 figure 中显示的结果时，可以用现在时。

Table 4 *shows* that growth was dependent on temperature.

计算的结果和统计分析结果应该用现在时。

The calculated value *is* significantly lower indicating most of the dissolved compound was degraded.

关于时态的用法没有严格的规定。现在越来越多的作者除了实验部分用过去时，其他部分都用现在时。美国化学会的《ACS Style Guide》建议实验部分用过去时，其他部分可以用过去时，也可以用现在时，但要保持一致。

3.1.10　主动句和被动句

许多人认为科技论文都应该用被动句，不要加入个性的成分。现在越来越多的杂志提倡使用主动句，因为主动句更简洁和明确。把"it is reported by the authors of this paper that"改为"we reported that"就显得更简要和直接，下面是个例子：

In 2002 we reported the synthesis of anthramycin analogues and their DNA binding activities studied by gel electrophoresis.

但实验部分还是主要使用被动句，用"The drug concentration was measured by HPLC"，而一般不用"We measured the drug concentration by HPLC"。

3.1.11　标点符号的使用

英文科技论文写作中经常使用的标点符号有逗号、句号和分号，冒号和问号使用的情况很少，而惊叹号几乎就不会被使用。现在分号的使用也逐渐减少，一般用句号取代。一篇论文中只是使用逗号和句号也是正常的。句号的使用比较明确，下面主要对逗号的使用作一些说明。逗号虽然很小，但要表达清楚你要传达的信息却离不开它。比如下面的例子，没有逗号时句子不好读，加逗号后，逗号放在不同的地方，意思完全不同。

Although it was incubated at 50 ℃ for 24 hours no reaction occurred.

Although it was incubated at 50 ℃ for 24 hours, no reaction occurred.

Although it was incubated at 50 ℃, for 24 hours no reaction occurred.

逗号是用来分开两个独立的句子，下面的句子是可以的。

Ethanol is used to replace gasoline, and it is produced from corn.

下面的句子就不合适，因为逗号后面不是一个完整的句子。

Ethanol is used to replace gasoline, and is produced from corn.

当一个长的语句出现在句子的前面时，要用逗号分开。若语句不长，不需要停顿，也可以不用逗号。

During the process of solvent removal, some crystalline solid was formed.

During the process precipitate was formed.

用分号时，分号后面的单词第一个字母要小写。

Duplicate simulations were performed at each temperature starting with structures found at different time points of equilibration; they showed a high degree of reproducibility with similar structure characteristics and order of unfolding events.

3.1.12 数字的写法

科技论文离不开数字。数字的一种写法是用英文字，如 three，thirteen。另一种是写阿拉伯数字。具体用哪种写法可参照以下几个简单的规则。

（1）少于 10 的整数用英文字，大于或等于 10 的数字用阿拉伯数字

three experiments

one assay

23 birds

6,500 miles

（2）有小数点和单位的数字用阿拉伯数字

1.2 hours

5 percent

3 am

page 3

（3）在句子开头的数字用英文字

英语句子的开始用大写字母来表明时，如果是一个数字，那就不能起到表示

一个句子开始的作用，所以不能用数字开始一个句子。这时要把数字用英文字写出，或最好能修改句子，不用数字开头。一个经常使用的办法是把表示数量的数字放到括号中去。

10 mL ethanol was added.

Ten mL ethanol was added.

宜改为：Ethanol (10 mL) was added.

30 eggs were used daily during the study.

宜改为：During the study，30 eggs were used daily.

（4）当两个数字前后并列出现时，一个要用英文字

当两个数字前后并列出现时，若都写成数字或英文容易产生混乱。

three eight-rat groups　to：three 8-rat groups

3 8-rat groups　　　　to：three 8-rat groups

two 5-day study

12 two-engine airplanes

（5）小于 1 的数字的单位用单数，大于 1 的数字的单位用复数

0.25 gram, 0.8 second, 1.5 grams, 3.45 seconds。

但是零后面的单位用复数。

zero meters, 0 meters。

单位的缩写单数和复数是一样的。

0.1 mL, 15 mL。

3.1.13　冠词的使用

使用冠词是英语的特点，中文没有相应字词。对定冠词 the 和不定冠词 a、an 的使用往往掌握不好，最常见的是漏掉。

There has been increase in loss of agricultural land.

宜改为：There has been *an* increase in *the* loss of agricultural land.

Stresses at various locations in crank are calculated by using sets of unit load cases applied to single throw FE model.

宜改为：Stresses at various locations in *the* crank are calculated by using sets of unit load cases applied to *a* single throw FE model.

再就是定冠词 the 用多了，例如：

The alcohol is produced by the fermentation of the grains like corn and wheat.

宜改为：Alcohol is produced by the fermentation of grains like corn and wheat.

alcohol 和 grains 都是泛指，不需要加定冠词 the。

除了第一个词，标题中的冠词不要大写。例如：

The Dependence of Crystal Growth on *the* Solvents.

3.1.14 同位词的使用

写作中有时需要对新引述的事物或概念做简单解释和描述，但单写一个句子又会打断前后的连接，这时用同位词来解释是一个经常使用的方式。

The Hallervorden-Spatz syndrome, *a neurological disorder associated with iron accumulation*, has been linked to a decline in cysteine dioxygenase activity.

同位词应该与同修饰的事物等位，说明是什么，而不能用来解释要修饰的事物的性质。下面句子中的同位词的使用就不合适。

Penicillin, not stable in water, was developed during WWII.

同位词要用逗号分开。

A novel compound, cyclic hexamethylidyne has been synthesized.

宜改为：A novel compound, cyclic hexamethylidyne, has been synthesized.

3.1.15 多余的用词

The supernatant was collected and concentrated in vacuum to evaporate the solvent.

concentrate 就是 evaporate the solvent，后面的 to evaporate the solvent 是多余的。把一些词的含义再次重复说明是最容易出现的多余用词。类似地，8 p.m. in the evening，blue in color，small in size，in vivo animal models，in vitro cell cultures 都有多余用词的情况。

The results of activities of tested compounds are listed in Table 1.

results 是多余的。中文经常写作"实验结果"、"活性检测结果"，但英文中"结果"是不需要翻译出来的。

宜改为：The activities of tested compounds are listed in Table 1.

下面的例句中，their 就是 acid B and D，二者要去掉一个。

The chemical mechanisms of their antioxidant activities of acid B and D were not well understood.

宜改为：The chemical mechanisms of the antioxidant activities of acid B and D were not well understood.

一些词组，如 it should be mentioned，it should be pointed out，it was found，it was determined……都是冗长和委婉的说法，应该直接说明。

3.1.16 隔离

英语中有些字是必须连在一起使用的，不能被隔离。

the almost same activity 应改为：almost the same activity.

have extensively been studied 应改为：have been extensively studied or have been studied extensively.

to carefully examine the structure activity relationship 应改为：to examine the structure activity relationship carefully.

3.1.17 Units

单位的写法上常出现的问题是数字和单位之间不用空格，如：3.3 mg not 3.3mg，4.0 mL not 4.0mL；体积 liter 缩写成 L，不用小写 l。

① Symbols or abbreviations do not have a period (.) unless at the end of a sentence.

② Small letters (lower case) are generally used, except for symbols derived from the name of a person. Exceptions are liter, which is L, and molarity, which is M.

③ The singular and plural forms of a symbol are the same. For example, 2 g not 2 gs, 2 mL not 2 mLs.

④ A space separates the number and the symbol, like 1.2 s, 45 mM, 37 ℃. Exceptions are the symbols for plane angular degrees, minutes and seconds (°, ′ and ″). There will be no space between, like 45°, 5′, 23″.

⑤ Symbols for derived units formed from multiple units by multiplication are joined with a space or center dot (·), e.g. "N m" or "N·m".

⑥ Symbols formed by division of two units are joined with a slash or given as a negative exponent. For example, the "meter per second" can be written "m/s", "m s^{-1}", "m·s^{-1}" or a slash should not be used if the result is ambiguous, i.e. "kg·m^{-1}·s^{-2}" is preferable to "kg/m·s^2".

⑦ Some commonly used units

kg, g, mg, μg, ng, pg

L, mL, μL, nL

mg/mL, μg/mL, ng/mL; better not use g/L, mg/L

mM, μM, nM, better not use mmol/L, μmol/L

M for molar, mol for mole

h for hours, min for minutes, s for seconds, m for meter

m, cm, mm

A, K, Hz, J, W, V, N, Ω

3.2 句子的连接和信息的传承

即使每个句子的语法合理，用词准确，但也不能保证论文就通畅。文字只是信息传递和描述逻辑思维的工具，文字用不好，当然信息传递就会出问题，但信息本身的传承，思维的逻辑性也很重要。文字上要把这个过程描述出来。我们先来看下面这个例子。

Some improvements of streptomycin fermentation have been achieved. For instance, the biosynthesis of streptomycin and its genetic control were described in a review, and one of its metabolic precursors was isolated. The batch-type feeding of carbohydrate substrates resulted in a yield increase of 23%–24%. The effect of carbon source consumption rate on streptomycin production was reported, and the yield was found to be about 2.0-fold higher than that of starch medium after 34 g·L^{-1} of olive oil was used as the sole carbon source.

在这个例子中，作者试图阐述 streptomycin 发酵生产改进方面的文献研究。作者以一句综述开头，这种开头的方法是很好的。然后列了四个方面的改进研究：

① biosynthesis and genetic control

② metabolic precursor

③ batch-type feeding

④ carbon source

作者只是把这四个方面写了出来，但没有连接关系，读起来就很不连贯。事实是列出来了，但没有有机地衔接起来。在改进的段落中，用时间关系把①和②来连接，用了 some times ago, recently。①-②和③-④的连接用了 "In addition to biosynthetic pathways, fermentation conditions were studied to improve production yield." 这句话。从一个状况到另一个不相关的状况时，用 in addition to 的句子是一个好的衔接办法。③和④之间的连接使用了连接副词 moreover。

Great efforts were made to improve the fermentation yield of streptomycin. Some of its major metabolic precursors were isolated and identified *some times ago*. *Recently*, the biosynthesis pathways of streptomycin and its genetic control were described. *In addition to* biosynthetic pathways, fermentation conditions were studied to improve production yield. The use of batch-type feeding of carbohydrates resulted in an increase of yield by 23%–24%. *Moreover*, the effects of carbon source consumption rate on streptomycin production were investigated. The yield when 34 g/L of olive oil was used as the sole carbon source is about 2.0-fold higher than when starch medium was used.

文章的紧凑和连贯要注意两个方面：一是内容，也可以说信息要连贯；二是用好连接词和连接用的副词。

3.2.1 信息的传承

信息或内容的衔接就是要求每个句子中的主题（主语）在前面的一个句子中要交代过。如果一个新的内容猛然间出现，就会打乱思绪，有不知为何的感觉。

There is increasing interest in *natural antioxidant* products for use as medicines and food additives. Vitamin C, vitamin E and carotenoids are some of these widely used *natural antioxidants*. *Reactive oxygen species* (ROS) including superoxide anion radical, hydroxyl radical and hydrogen peroxide, are generated under physiological and pathological stresses in human body.

这个例子中，第一句和第二句都与 natural antioxidant 有关，但第三句出现了 reactive oxygen species，与前面的内容没什么关系，这时第三句与前面的传承上出现了断离。下面的改动中加了个连接句子，把 natural antioxidant 和 reactive oxygen species 连接了起来。

There is increasing interest in *natural antioxidant* products for use as medicines and food additives. Vitamin C, vitamin E and carotenoids are some of these widely used *natural antioxidants.Antioxidants* played an important role in lowering oxidative stresses caused by *reactive oxygen species* (ROS). *ROS* including superoxide anion radical, hydroxyl radical and hydrogen peroxide are generated under physiological and pathological stresses in human body.

再看下面的例句。

Streptomyces species were widely used to produce *antibiotics*. *Streptolydigin* was a tertramic acid antibiotic, which was produced through polyketide pathway by *Streptomyces lydicus.*

这个例子中第一句的 *Streptomyces* 和 *antibiotics* 与第二句的 *Streptolydigin* 关系不清楚，第一句中应该有 *Streptolydigin* 出现。在改动的句子中，一个举例的接词 such as 把 *Streptolydigin* 引出。当以综述开始写作时（general to specific），用举例的方式把要研究的特例引出是一个经常使用的方式。

宜改为：*Streptomyces* species were widely used to produce various *antibiotics*, such as *streptolydigin*. *Streptolydigin* was a tertramic acid antibiotic that was produced through the polyketide pathway by *Streptomyces lydicus*.

从一个状况到另一个不相关的状况时，用 in addition to 的句子衔接。

In addition to the kinase activity, it also showed a peptidase activity.

Oxygenases catalyze the incorporation of a molecule of oxygen (O_2) into the substrate. They catalyze the initial and rate-limiting step of L-Trp catabolism in the kynurenine (Kyn) pathway ⋯ *In addition to* its role as a L-Trp-catabolizing enzyme, IDO is involved in the immuno-regulating system ⋯

3.2.2 句子的连接

科技论文有很强的推理特性。要表达一个思维过程，思想之间的传承和衔接要紧密、合理，思维是用句子表达出来的，这样句子之间的衔接和关系就特别重要。句子的衔接可以通过连接词、短句或一个整句，衔接的作用就是把前后的思想贯穿起来，从而达到说明、推理、讨论的目的。

句子可以分为简单句和复杂句。复杂句中的短语之间需要用连接词和副词来连接，同时句子之间的传承也需要使用连接词和副词。如 and，but，or，for，nor，yet，not only ⋯ but also ⋯，either ⋯ or ⋯，although，after，before，because，if，as， when，since，than，where，unless，though，whereas 都是些经常使用的连接词。

除此之外，科技写作中经常使用一些连接副词，如 also，however，moreover，furthermore，therefore，otherwise，consequently，indeed，similarly，finally，likewise，then，hence，nevertheless，thus。

当用连接副词来连接两个独立的句子时，一般语法书中都注明第一个句子后加分号，连接词后加逗号。但现代杂志论文中往往在第一个句子后加句号，而不用分号，两种写法都是可以接受的。

We quantified the extent of forest damage in the region; *moreover*, we attempted to explain why trees died.

We quantified the extent of forest damage in the region. *Moreover*, we attempted to explain why trees died.

It was rained; *however*, the drought persisted.

It was rained. *However*, the drought persisted.

根据前后连接的关系，连接词和副词可分为 7 大类。

（1）结果关系

经常使用的词有 therefore，consequently，thus，hence，as a result，indeed 等。

If gasoline is to burn inside the engine, oxygen must be present to support the combustion. *Therefore*, the fuel should be a mixture of gasoline and air, which contains oxygen.

Climate change is expected to have an impact on the timing of sprouting of plants living in seasonal environments. *Indeed*, analysis of long-term phenological data has revealed significant trends linked to climate signals, such as temperature.

（2）同等关系

当描述同等的事物或特性时，如果没有用来连接的字词，往往有罗列和前后关系不明的感觉。furthermore，moreover，also，in addition，besides 等连接副词的使用使句子更通畅。

also，too，likewise 表示新引进的事物与前面的同等重要，只是在语气上有所不同。too 比较随便，日常用语中使用较多。also 要正式些，likewise 非常正式。科技写作中，also 用得最多，likewise 也可以。besides 引入的事物多用来补充和加强前面的声明。in addition，moreover 和 furthermore 多用于强调新引进的事物的重要性。

Chlorins play important biological roles. They inhibit certain oxidative stress in some marine species, including sponge, clam, and scallop. *Also*, one of the chlorins is a hormone responsible for sexual development in a marine worm. *Moreover*, in addition to their natural functions, they showed potential applications in medical and material sciences.

This observation suggests that the origin of this effect is not purely of an acoustic nature. *Furthermore*, the instability is not a Helmholtz-type oscillation, because the frequency shows little response to variation of the combustion chamber or plenum geometry.

The experimental conditions we found have not been explored by any previous work. *In addition*, no theoretical calculations have predicted the instability of the alloys.

（3）相反关系

论文写作中，当描写一些与设想和文献不符的现象或结果时，需要使用 in contrast，but，however，yet，on the other hand，surprisingly，nevertheless，instead 等表示转折关系的副词连接。

The chlorins are a class of compounds made of aromatic tetrapyrroles. The most important one is chlorophyll, which is used as a light-harvesting chromophore in photosynthetic species. *However*, many lesser known chlorins also play important biological roles.

It is difficult to attribute differences in decay constants to specific properties of the bridge. *Instead*, we can characterize electron tunneling systems in terms of effective barrier height.

（4）相似关系

引出有同等和类似关系的事物时，用 similarly，likewise 等。

We found that elevated CO_2 in the atmosphere only causes accumulation of soil carbon content when nitrogen is abundant. *Similarly*, elevated CO_2 only enhances N_2 fixation when other nutrients, such as phosphorus and potassium, are added. Hence, soil C sequestration under elevated CO_2 is constrained both directly by N availability and indirectly by nutrients needed to support N_2 fixation.

（5）举例

从综合的阐述到特定例子的写作方法在科技论文写作中经常使用。用举例的方式引出自己的研究对象，这时需要如 for example，for instance，specifically，such as，including 等词。

Many gastro-intestinal peptides, *including* secretin, PPY, and ghrelin, have receptors in the brain.

The population growth in the US is affected by many factors, *such as* race, religion, income, and cultures. *For instance*, Jewish family has an average birth rate of 3.4 children per couple and Caucasian white family has a birth rate of 1.8 children per couple.

（6）时间

用于时间关系连接的词有 since then，after that，now，later 等。

It is generally believed that an infectious agent, such as a bacteria or a virus, has to have nucleic acids as their genetic code so it can multiply in the host. *Now*, the studies of prion have demonstrated that a protein can also act as an infectious agent and multiply in the host to cause diseases.

（7）顺序

用于前后顺序的连接有 then，next，finally，first，second 等。

In this study, we report that G protein promotes lung cancer cell invasion. Moreover, we demonstrate that inhibition of G protein reduces the metastasis of lung cancer cells *in vivo*. *Finally*, we demonstrate that the expression of G protein is significantly up-regulated in the early stages of lung cancer.

The microplate was shaken for 5 minutes at room temperature and *then* incubated at 4 ℃ for 24 hours.

3.2.3 平行句的组织方法

当一步一步地描写实验操作时，可以把几个动作合并到一个句子中，这样紧凑，内容也清晰。这时一个句子中有几个并列的动词，当只有两个动词时，用 and 连接；当有三个以上的动词时，每个动词用逗号分开，最后一个动词前面加 and 连接，and 前加逗号。组织这样的句子时要注意主语和并列的动词之间的一致性。下面是几个例句。

The combined extract was filtrated and then condensed at 50−60 ℃. The condensed extract was precipitated with ethanol (final concentration at 70%) over night, and then the supernatant was collected, concentrated in vacuum (40 ℃) to evaporate the solvent.

宜改为：The combined aqueous extract was *filtrated*, *condensed* at 50−60 ℃, and *precipitated* with ethanol (to a final ethanol concentration of 70%) over night. The supernatant was then *collected* and *concentrated* in vacuum (40 ℃).

Streptolydigin was extracted from the supernatant fluid by adding ethyl acetate with the ratio 1:1. The extracts were combined, and the solvents were concentrated under reduced pressure, and the oily residues were redissolved in 5 mL of methanol. Then the methanol solution of extracts was filtered through a 0.45 μm syringe filter for HPLC analysis and was stored at −20 ℃ until being analyzed.

宜改为：Streptolydigin was extracted from the liquid layer with equal volumes of ethyl acetate. The extracts were then *combined* and *concentrated* under reduced pressure. The oily residue was *redissolved* in 5 mL of methanol, *filtered* through a 0.45 μm syringe filter, and *stored* at −20 ℃ until being analyzed.

下面再列出几个在 Materials and Methods 中提到的几个平行句的例子：

After 4 days of treatment the animals were *pulsed* with bromodeoxythymidine before sacrifice, *euthanized*, tumors *removed*, and *fixed* in 4% paraformaldehyde overnight and then *embedded* in paraffin.

20 μL samples were *removed*, *precipitated* with 60 μL acetone, *centrifuged* at 4 ℃ for 10 minutes at 3000 rpm, and *transferred* to a clean vial.

The combined organic layers were *washed* with water, *washed* with brine, *dried* over sodium sulfate, *filtered*, *concentrated* and *purified* on silica to yield 0.8 g of a white solid.

3.2.4 写简单句子

科技写作最主要的是要清晰、简洁、直接，要写简单的句子。句子长了，一是不容易懂，二是易犯语法错误。一位哈佛大学的编辑写道：

"American academic writing, for most subjects, is linear, succinct, and direct. Some international students have called American writing 'simple'."

对长句子，最好缩短成两个短句子。下面举些例子：

It has been well established that monoubiquitination of FANCD2 constitutes a key step in the FA-BRCA DNA repair pathway, and can be induced by various genotoxic stress including UV light, ionizing radiation, hydroxyurea, and cross-linking agents.

宜改为：It has been well established that monoubiquitination of FANCD2 constitutes a key step in the FA-BRCA DNA repair pathway. Monoubiquitination of FANCD2 can be induced by various genotoxic stresses including UV light, ionizing radiation, hydroxyurea, and cross-linking agents.

Porcine postweaning diarrhea (PWD) and edema disease (ED), associated with F18 fimbrial *E.coli* strains which adhere to and colonize the microvilli of small intestinal epithelial cells by F18 fimbriae and secrete Shiga-like toxin (Imberechts et al., 1994b; Imberechts et al., 1994a), are the mostly widespread causes of death in weaned pigs and account for substantial economical losses in swine factory (Gaastra and Svennerholm, 1996; Fairbrother et al., 2005).

宜改为：Porcine postweaning diarrhea (PWD) and edema disease (ED) are the major causes of death in weaned pigs and account for substantial economical losses in swine industry (Gaastra and Svennerholm, 1996; Fairbrother et al., 2005). PWD is caused by infections of F18 fimbrial *E.coli* strains that adhere to and colonize on the microvilli of small intestinal epithelial cells and secrete Shiga-like toxins (Imberechts et al., 1994b; Imberechts et al., 1994a).

Duplicate simulations were performed at each temperature starting with structures found at different time points for the equilibration and showed a high degree of reproducibility with similar characteristics and the order of unfolding events, while some differences in the time scale over which these events took place.

宜改为：Duplicate simulations were performed at each temperature starting with structures found at different time points of equilibration. They showed a high degree of reproducibility with similar structure characteristics and order of unfolding events. However, there were some differences in the time scale over which these events took place.

3-Methylindole is a compound found commonly in water containing decaying leaves and isolated from fermented Bermuda grass infusions, and known to be oviposition attractant for *Cx.quinquefasciatus* in laboratory bioassays (Millar et al., 1992; Mordue (Luntz) et al., 1992; Blackwell et al., 1993)and mediated oviposition of *Cx.tarsalis* and *Cx.stigmatosoma* under field conditions (Beehler et al., 1994).

宜改为：3-Methylindole is a compound found commonly in water containing decaying leaves. It was isolated from fermented Bermuda grass infusions and known to be an oviposition attractant for *Cx.quinquefasciatus* in laboratory bioassays (Millar et al., 1992; Mordue (Luntz) et al., 1992; Blackwell et al., 1993). It mediated oviposition of *Cx. tarsalis* and *Cx. stigmatosoma* under field conditions (Beehler et al., 1994).

We concluded that FANCD2 monoubiquitination might be regulated by different ubiquitin-related enzymes, which probably include UBE2T and UBE2W, in response to different types of DNA damage agents, and provide an additional regulatory step in the activation of the FA pathway.

宜改为：We concluded that FANCD2 monoubiquitination might be regulated by different ubiquitin-related enzymes, which probably include UBE2T and UBE2W, in response to different types of DNA damage agents. FANCD2 monoubiquitination provide an additional regulatory step in the activation of the FA pathway.

3.3 中式英文

由于英文与中文表达方式和语法结构的不同，直接翻译往往会写出不够母语化的句子。直接翻译的句子在语法上也许是正确的，但母语并不这样表达，按字面意思直接翻译有时会很可笑。要避免字面翻译，平时阅读时要多留心和注意母语的写法。例如，日常生活中我们常见到以下中式英文：

大雨，big rain，宜改为 heavy rain。

保持距离，Confirming distance，宜改为 Don't tailgate。

进入京张高速，to put in Jingzhang Expressway，宜改为 to Jingzhang Expressway。

欢迎你来北京，Welcome you to Beijing，宜改为 Welcome to Beijing。

学院路 18 号，Xueyuan Road No. 18，宜改为 18 Xueyuan Road。

小心滑倒，Carefully slip 或 Slip carefully，宜改为 Be careful not to slip and fall。

严防超速行驶，No overspeed driving，宜改为 No driving over the speed limit。

小心碰头，Be careful your head, Carefulness bump head, Carefully Bang head，宜改为 Mind your head。

类似的情况在论文写作中也经常出现。但中式英语还是有规律可循的，容易写成中式英语的情况可以概括为以下几种类型。

（1）用词不准确

由于词汇不足和对英语写法的不熟悉，一些简单的日常用词用得过多。这些简单的词一般都少于 4 个字母，应尽量避免使用这些简单词汇，而用较正式的学术词汇。

big (large，多、大)
drop (decrease，降低)
get (obtain，得到)
give (produce, yield, afford，得到)
help (contribute，有助于)
see (show，显示)
work (investigation，研究) etc.

do (carry out, perform，做)
for (because，因为)
get up (increase，增加)
go up (increase，增加)
keep (maintain, remain，保持)
test (investigate, examine，研究、检查)

一些写法上相似，但意义不同的词经常被混淆使用：

absorption 吸收 / adsorption 吸附
access 进阶 / assess 调查
accidental 偶然的 / coincidence 巧合
aspire 渴望 / aspirate 吸气
blockade 封锁 / blockage 堵塞
cheap 便宜的 / low-cost 低成本
character 人物特点 / characteristic 特征
close 临近 / similar 相似
conserved 保留，不消耗 / conservative 保守的
deduct 减掉 / deduce 推断
detect 检测 / measure 测量
design 设计 / designate 指定
diminish 递减，渐渐消失 / decrease 减少
dramatically 急剧地 / drastically 剧烈地
efficiency 效率 / efficacy 疗效
indicate 表明 / indict 起诉
induced 诱导 / provoked 挑起
increment 递增 / increase 增加
keep 保持 / maintain 维持 / remain 保持没有变 / retain 保留

object 物体 / objective 目的

manuscript 稿件 / paper（一般）论文/ article（综合性）论文/ report 报道

microscope 显微镜/ microscopy 显微镜技术

middle 中间 / central 中心，很重要

preceding 以前/proceeding 进行中

protect 保护 / preserve 保留

principle 原则/principal 校长、本金

present 介绍 / represent 代表

symptoms 症状 / syndrome 综合征

synergic 协同的/ combined 结合的

systemic 系统的 / systematic 系统的，详细的

shed 撒，散出 / shred 撕碎

transplant 移植 / implant 植入 / graft 嫁接

variation 变动 / change 变化，改变

underlining 标出（以示重要）/ underlying 潜在的（原因）

use 用 / apply 运用

wild 荒野 / wild-type 野生型

wildly 疯狂地 / widely 广泛地，普遍地

wrong 错误的 / false 虚假的

（2）多余用词

从中文翻译成英文时，往往会出现一些多余的用词，也就是说一些中文字词是不需要翻译出来的。这与中英文语言特点有很大关系。中文为了朗朗上口，会有一些词语重复表达，例如"意见和建议"；有的会出现"动词+动词名词化"，如"进行探索"；以及多余的修饰词，如"相互合作"。但英文中用一个词即可表示,例如,意见和建议用suggestion、进行探索用explore、相互合作用cooperation等。

"The antibody cost is expensive" 抗体的价格高昂，"价格"不需要翻译出来，宜改为："The antibody is expensive"。

"The results of the HPLC analysis showed that…" 液相色谱分析结果表明，"结

果"不需要翻译出来，宜改为："The HPLC analysis showed that…"。

"Triple combination of A, B, C"，A, B, C 三组合，"三"不需要翻译出来，宜改为："Combination of A, B, C"。

"The length of the amino acid sequence of rcl is 578"氨基酸顺序长度是 578，"长度"不需要翻译出来，宜改为："The amino acid sequence of rcl is 578"。

"The amount of reduction" 减少值，"值"不需要翻译出来，宜改为："The reduction"。

"The pH value was measured" 检测 pH 值，"值"不需要翻译出来，宜改为："The pH was measured"。

（3）词组的顺序

中文的用词顺序和英文有不同之处，直接翻译会写出怪异的句子，这是最经常出现的中式写法，写作时一定要注意。

"Locate at JiLin province Songhua River branch Mengliu River recently happen chemical pollution incident." 这是中文 "位于吉林省的松花江支流孟柳河最近发生了化学污染事故"的直接翻译，与英文的顺序不符。

宜改为："A pollution incident happened recently in Mengliu River, a branch of Songhua River located in Jilin Province."

"According to understanding, incident pollution reason already basically investigates clear, by JiLi Changbai Mountain Chemical Limited Corp. towards Mengliu River in deliberately release chemical waste water cause." 这也是中文"据了解，污染的原因已基本明确，是由……" 的直译，翻译成英文时，有些不需翻译，有些字序要改变。

宜改为："Investigations have revealed that the cause of the pollution incident was JiLi Changbai Mountain Chemical Limited Corp. deliberately released chemical waste into the Mengliu River."

"Nod factor sulfate modification"，"硫酸化修饰"的直译是"sulfate modification"，意译是"modification by sulfate"。

宜改为："Nod factor modified by sulfate" or "modification of Nod factor by sulfate"。

"Closed-ring form"，"闭环"的直译。

宜改为："Ring-closed form"。

"Synchronous multiple hepatic metastases"，"同步多发肝转移"的直译。

宜改为："multiple synchronous hepatic metastases"。

"Pharmacological complete response""药理完全响应"的直译。

宜改为："Complete pharmacological response"。

"Specific antibody to Sudan red was prepared","苏丹红的特异抗体"的直译，特异抗体不是 specific antibody，而是 antibody specific to。

宜改为："Antibody specific to Sudan red was prepared"。

"Significant regions of activation were listed","显著的活化区域"不是 significant region of activation，而是 region of significant activation。

宜改为："Regions of significant activation were listed"。

"Functional magnetic resonance imaging (fMRI) Researches of the brain semantic memory""核磁共振研究脑记忆"的直译，有些字序要改变。

宜改为："Researches of the brain semantic memory by functional magnetic resonance imaging (fMRI)"

"Wanshou Road 24"，地址的直译。

宜改为："24 Wanshou Road"。

"Docetaxel new formulation"为"紫杉醇新制剂"的字面翻译。

宜改为："new formulation of docetaxel"

"Primary hydroxyl groups free saccharides","含有一级羟基基团的多糖"的直译。

宜改为："saccharides with free primary hydroxyl groups"。

"We concluded that H_2S and NO have a stimulating capillary proliferation effect","stimulate capillary proliferation effect"是"刺激血管增长效果"的直译，应意译为"stimulate the proliferation of capillary"。

宜改为："We concluded that H_2S and NO stimulated the proliferation of capillary"

（4）主句和修饰句的位置

中文的语法特点是先说修饰部分，最后说主句。比如:今天天气不好，我就没去学校。当温度提升到100℃时，反应的产率增加到92%。这两句话直译成英文是：Because of the bad weather, I did not go to school. When the temperature was raised to 100℃, the reaction yield increased to 92%.

然而，英文则相反，先说主句，后面跟随修饰部分。上两句话翻译成标准英文是：I did not go to school because of the bad weather. The reaction yield increased to

92% when the temperature was raised to 100℃. 英文习惯是先把主句，也就是作者的观点说出来，后面再加各种修饰条件。中文则相反，表达的观点被放在各种修饰句子后面。这与西方文化直率，直入主题，而中国文化委婉，慢慢道来，有些相似。

如果想强调条件，比如天气不好，把它放在前面也是可以的。

下面再列举一些在论文写作中类似的中式英文。

In SS and SS–13BN rats, we found that after high salt intervention, mRNA and protein level of renalase in kidney was significantly decreased (P<0.05), suggesting that salt may induce salt-sensitive hypertension through inhibition of renalase. 在 SS 和 SS-13BN 大鼠中，我们发现高盐干预后，肾脏中肾脏酶的 mRNA 和蛋白水平明显下降（P<0.05），表明盐可能通过抑制肾脏酶诱发盐敏感性高血压。

这段句子中把修饰条件"In SS and SS–13BN rats"和"after high salt intervention"放在前面，后面才说主要的对象"mRNA and protein level of renalase in kidney"。

宜改为：We found that mRNA and protein levels of renalase in the kidneys of SS and SS–13BN rats were significantly decreased ($P < 0.05$) after high salt intervention, suggesting that salt may induce salt-sensitive hypertension through inhibition of renalase expression.

To probe the surface chemical properties and the interaction between the constituents of catalysts, X-ray photoelectron spectroscopy (XPS) measurements were carried out. 为了探究催化剂的表面化学性质和各成分之间的相互作用，我们进行了 X 射线光电子能谱（XPS）测量。

中文一般说 "为……目的，做了……实验"，而英文先说实验，再说目的。

宜改为：X-ray photoelectron spectroscopy (XPS) measurements were carried out to probe the surface chemical properties and the interaction between the constituents of catalysts.

When catalyzed by 0.26% CoO/Al_2O_3, the highest CH_4 production rate of 189 mmol·g_{cat}^{-1}·h^{-1} was obtained. 当由 0.26%的 CoO/Al_2O_3 催化时，获得了最高的 CH_4 反应速率，即 189 mmol·g_{cat}^{-1}·h^{-1}。

中文先说条件（催化下），再说得到的结果。英文顺序相反。

宜改为：The highest CH_4 production rate of 189 mmol·g_{cat}^{-1}·h^{-1} was obtained when catalyzed by 0.26% CoO/Al_2O_3.

With the progression of kidney disease, the prevalence of anemia increases with more than 90% of dialysis patients having the condition. 随着肾脏疾病的发展，贫血

的发病率增加，90%以上的透析患者都有这种情况。

中文先说条件（病情恶化），再说得到的结果。英文则顺序相反。

宜改为：The prevalence of anemia increases with the progression of kidney disease and more than 90% of dialysis patients have the condition.

However, using several superior vascular ultrasound imaging techniques to evaluate the internal carotid artery dissection has not been reported. 然而，使用几种优越的血管超声成像技术来评估颈内动脉夹层的情况还没有报道。

中文一般说"用……方法，做……实验"，而英文说"做……实验，用……方法"。

宜改为: However, evaluation of the internal carotid artery dissection using several superior vascular ultrasound imaging techniques has not been reported.

（5）论文中经常出现的中式写法

中文论文中经常用"为……提供了理论依据"或"为……提供了科学依据"作为科研成果的意义，"依据"的英文直译是 basis，但英文不这么表达，常用 rationale。下面是几个例子：

"scienctific basis" 宜改为 "scientific rationale"。

"theoretical basis" 宜改为 "theoretical rationale"。

"experimental basis" 宜改为 "experimental support"。

"all the document" 全篇（修改），宜改为 "entire document"。

"discover" 用于表示新发现，一般论文中宜改为 "show, demonstrate"。

"normal"，通常(情况下)，正常(条件下)。不特指，不清楚具体指什么。"normal control" 宜改为 "negative control"，"normal patient" 宜改为 "healthy patient"。

"performance"，表现，性能。太笼统，英文中很少用，宜改为更具体的术语如 "capability"，"activity"，"reactivity" 等。

"situation"，情况，形势。太笼统，不清楚要说什么，宜改为更具体的术语如 "disease profile"，"conditions"，"phenomenon" 等。

"speed" 特指与距离有关的速度，其他的地方应该是速率，宜改为 "rate"。

"test the efficiency"，test 是检查，检测。测定或测量宜改为 "determine/measure the efficiency"。

第4章
常用词汇

写作时，最困难的是如何把想写的一句话准确地用英文表达出来。这个困难源于词汇的缺乏，也源于对一些词汇的准确意义掌握不准。而缺乏的词汇往往不是专业名词，而是一些动词、句间连接词和一些适当的修饰词。在日常写作中要特别注意积累一些自己专业内经常使用的一些词汇，建议读者建立自己专业写作的小字典，用心去学习每个关键词的正确使用方法。本书从大量的不同专业的文章中总结出 300 个左右经常使用的词汇介绍给读者，这些词汇为关键词。一篇论文大约有 3000～5000 字，考虑到重复使用，实际用的词汇大约在 1000～2000 字。除了普通的简单词汇和专业术语，也就有 200 个左右的动词、连接词和修饰词需要特别学习掌握。学会了这些关键词的使用，写起文章来就会很顺利，也就能把自己的科研结果和思想清楚地表达出来了。

为了便于整理和使用，这里把常用的词汇分成 12 组讲述。这个分类并不严格，只是为了整理和使用的方便。写作时要避免把汉字简单地直译为英文，希望读者能从例句和日常阅读中领悟词的含义，从而达到准确使用的目的。对字的使用应特别注意以下几点：

（1）用字要准确

每个科学术语都有其特定的含义，使用要准确。比如微生物学中经常使用 medium、broth 和 culture 三个词，但它们之间存在差别：medium 是培养基，可以是液体也可以是固体；broth 是培养液，只是液体；而 culture 是指细菌和培养液的混合物。如果要从发酵罐中取样，那取的样应该是 culture，而不能用 medium 或 broth。化学中的量和浓度有明确定义，mole 和 molar 就大不一样。molar 和 mole 对有些化合物是一样的，对有些化合物则是不同的。物理学中的温度、距离、时间等均有自己的严格定义。科研人员对自己专业的词汇一般都能较好地掌握，但对一些普通词汇的使用常掌握不准。举几个简单例子：

比如，promote 一般指职位的提升，不能当 increase 来使用。

Production was <u>promoted</u> by 16% in the new procedure.

宜改为：Production was *increased* by 16% in the new procedure.

比如，increment 是指少量或一段小的变化，可以是增长，也可以是降低，与 increase 不同。

An <u>increment</u> of production by 16% was achieved.

宜改为：An *increase* of production by 16% was achieved.

再比如，perform 表示一个行动，而不指某个具体事物。浓度不能 perform，但测浓度可以用 perform。

The zinc concentration was <u>performed</u>.

宜改为：The zinc concentration was *measured*. 或 Measurement of the zinc concentration was *performed*.

下面再举一些用字不准的句子。

Few papers can be <u>seen</u> discussing the effects of precursor feeding on the antibiotic production.

宜改为：Few papers were *published* discussing the effects of precursor feeding on the antibiotic production.

It could be <u>seen</u> from Table 1 that compound 12 is the most active one.

宜改为：As *shown* in Table 1, compound 12 is the most active one.

<u>Relative to</u> peptide Y, peptide NP was less studied.

宜改为：*Comparing with* peptide Y, peptide NP was less studied.

The protective activities of Sal B and Sal Y were significantly <u>distinct</u>.

宜改为：The protective activities of Sal B and Sal Y were significantly *different*.

These results were <u>consensus</u> with those previously reported data.

宜改为：These results were *consistent* with those previously reported data.

（2）推理用语的使用

从实验观察和数据到结论的推理过程中，在事实和理论的关系上可能有从"与……一致"到"表示"到"证明"等不同的强弱关系，在选择用词上要合理。英语中经常使用的词是 is compatible with，imply，suggest，indicate，show，prove，这基本上是一个从弱到强的顺序。

is compatible with，is consistent with，in line with 表示是个合理的解释，不矛盾，但可能还有其他的解释。

imply，suggest 表示支持现在的结论，这个结论比其他的更合理，但不能证明就是这个结论。

indicate，show，demonstrate 就更确定，表示几乎就是这个结论了。其他的可能性不大，但还不是百分之百的证明。

prove 表示完全肯定，没有任何其他可能性了。科研中很少有这种情况，用

prove 要特别注意。

（3）用词要保持一致

论文从头到尾用字要保持一致。比如时间单位的分钟即可以全拼成 minutes，也可以简写成 min，二者都可以，但只能用一种写法。Figure 或 Fig.只用一种写法。数字的写法上，如果一个写成 fifteen，另一个写成 12 也是不一致。修改论文稿件时应特别注意用词的一致性。

（4）尽量少用简写

作者自己熟悉的简写，读者不一定知道，所以应尽量少用简写。必需时，宜于第一次出现全名时在其后用括号标出简写。比如 Heat Shock Protein (HSP)、Multi Body Dynamics (MBD)。至于广泛使用的简写，如 mL、AIDS 等可不用写出全名。

（5）不要用缩约语

正式写作中，不要使用如 didn't，don't，can't，haven't……缩略语，应该写完整形式，did not，do not，cannot（一个字），have not……

（6）避免使用俚语

虽然非英语国家的作者不能很好区分什么是俚语，但还是要注意不要使用像 a lot，sort of，pretty good……的口头用语。

4.1　经常使用但容易出现问题的字

above

经常用来指前面提到过，表示"如前所述，as discussed above"、"前面的方法，the above method"等。但 above 不确定时，用起来容易，读起来不易明白，容易造成表达不清。这时应把所指的事情明确地写出来。类似不明确的字还有 former，latter，应避免使用。

absorption，adsorption

adsorption is to the surface, not inside.

access，assess

access 是接近；assess 是评估，估价。

adapt, adept, adopt

adapt 是动词，适应；adept 是形容词，熟练的；adopt 是动词，采用。

affect, effect, impact

affect 是动词，影响（influence）的意思。如"温度对细胞生长的影响"可表示为"Temperature affected the growth of cells"。

effect 是名词，是结果（result）影响的意思。有时也可以作动词，招致（bring about）的意思。科技论文中很少使用 effect 的动词形式。

impact 当冲击，碰撞讲。如"Western popular culture has a huge impact on Asian society"。自然科学不同参数的相互影响一般不用 impact 来描写，用 affect 或 influence 更合适。

agree to, agree with

agree to 是同意，giving consent，"I agree to a biopsy test"；agree with 是一致，in accord，"The results agree with our previous observation"。

alternate, alternative

alternate 是交替的，轮流的；alternative 是另外的，选择的。

and

and 是一个非常有用的词。它是一个最被经常用到的连接词，可以连接类似的词、词组或句子。用 and 连接的词、词组或句子是相关联但又各自独立的。当排列三个以上的词时，最后一个词之前加 and，其他的词后都加逗号。

Pollution in the river affected the population of different animals, such as fishes, birds, *and* turtles.

and 之前也可以不加逗号。

Pollution in the river affected the population of different animals, such as fishes, birds *and* turtles.

一般认为 and 之前加逗号是美国写法，不加逗号是英国写法。

用 and 连接句子时，若两个句子都很简单，中间可以不加逗号。

One liter of water was added *and* the solution was left at 4 ℃ overnight.

但中间加逗号也正确。

One liter of water was added, *and* the solution was left at 4 ℃ overnight.

两个比较复杂的句子中间一定要加逗号，以便于阅读和理解。

The assay was carried out by heating the sample in boiling water for two hours, *and* the volume of the assay solution was kept constant by adding water.

用 and 来开始一个句子也可以，and 起到对一些相关和并列的描述来连接的作用。

AIDS drugs are effective to control the replication of HIV, but they cannot cure AIDS. *And* a vaccine for AIDS is still elusive. Therefore, education and prevention are the most effective weapons against AIDS for now.

用 and 连接两个独立的句子时，可以在 and 之前加逗号，也可以不加逗号。但连接一个不完整的句子时，不能用逗号分开。

The sample was added to the testing solution, <u>and</u> allowed to react in a water bath at 37 ℃ for 10 min.

宜改为：The sample was added to the testing solution *and* allowed to react in a water bath at 37 ℃ for 10 min.

but 的用法与 and 很类似，只是它连接的句子是对照和相反的。

apparent (apparently)

apparent 有"明显的，obvious，clear"和"貌似的，seeming"两种用法，具体是哪种用法，句子的前后内容可能会有提示。在不能确定的情况下，应避免使用，选用 obvious 或 seeming 更好。

apparently 也是一样，选用 obviously 或 seemingly 更好。

appear

appear 有"出现，to come into view"和"好像，seem"两种讲法，一般在科技论文中，当"好像，seem"的用法要多些。

After cooling down to room temperature, yellowish crystals *appeared*.

It *appears* that deformation of frogs in the Hui River is caused by pesticides.

as

as 有很多不同的讲法，它有表示因果关系的用途，类似 because 和 since。但 as 表示的因果关系最弱，since 在中间，because 的因果关系最明确。所以不要用 as 来表示因果关系，用 because 或 since。

as 用来表示"同样地","像……一样"。

The genetically engineered apple tastes *as good as* the natural one.

As soon as the body temperature reached 39 ℃, the child has to be sent to the emergency room.

也当"在……的时候，when"讲。

As the snow started to melt, the team went back into the forest to collect samples.

augment

to make (something already developed or well under way) greater, as in size, extent, or quantity. 继续增长，进一步扩大。

average，mean，median

average 和 mean 都是平均的意思，mean 是个数学用语，median 是一个系列中的中间的那个值。

如 2，4，8，16，32，average 和 mean 是 12.4，median 是 8。

because，because of

用 because 来表示因果关系是最明确的。Because 后面写句子，because of 后面用名词。

The tiger population decreased dramatically *because* their habitat was destroyed.

The tiger population decreased dramatically *because of* loss of habitat.

for 和 since 表示的因果关系要弱些，用 since 时往往强调的是当时的情况、时间、地点等，有中文"既然"的意思。as 表示的因果关系最弱。

Since the weather is hot and humid, we decided to take a break.

below

与 above 一样，所指含糊，应避免使用。

beside，besides

beside 是"在……旁边"，besides 是"除了……"。

between，among

between 是两个人或事之间，among 是两个以上人或事之间。

but

与 and 一样，but 是一个被经常使用的连接词，它连接的句子有对照和相反的意思。

在连接比较简单的两个句子，并不影响句子的流畅的时候，but 之前可以不加逗号。

He felt better *but* did not fully recover after taking the medicine.

连接两个比较长的句子时，but 前应加逗号，把两个句子分开。

Scientists spent months to figure out why the satellite did not reach its orbit, *but* they never find the truth.

but 也可以放在句子的开头。

For a long time people realize there has to be a natural ligand for the cannabinoid receptor. *But* its identity was only elucidated recently.

用逗号把 but 隔开是不对的，下面的用法就是错误的。

But, its identity was only elucidated recently.

cannot

cannot 是一个字，can not 应写成 cannot。

case

case 只是起到填充空间（filler）的作用，没有实质意义，应避免使用类似"in the case of"的词组。

compare with，compare to

compare with 是比较的意思，而 compare to 是比作的意思。

It is often to *compare* the human brain *with* a computer.

The human brain is often *compared to* a black box.

compose，consist，comprise

compose 当"组成，构成"讲，是及物动词。一般用"×××is composed of

×××"的形式。如：An atom is *composed* of a nucleus and a defined number of electrons.

consist 当"由…组成"讲，是不及物动词。一般用"××× consists of ×××"的形式。如：An atom *consists* of a nucleus and a defined number of electrons.

comprise 当"包括"讲，也有"由…组成"的意思。到底是什么用法，很容易弄不明白，所以最好不要用这个字。

conserved，conservative

conserved 保存的，保持的；conservative 保守的，守旧的。

continual，continuous

continual 是经常发生的；continuous 是连续和不间断的。

correlated with，correlated to

correlate with 是正确用法，correlate to 是错误用法，related to 是正确用法。

deduct，deduce

deduct 是扣除，deduce 是推理。

design，designate

design 是设计，构思；designate 是指定，指出。

different from，different than

事物之间不同用 different from，人物之间不同用 different than。具体的语境中 different from 不能写成 different than。

diminish，decrease

diminish 削弱，减为很少；decrease 减少，并不一定减为很少。

dramatically，drastically

dramatically 是显著的；drastically 是激烈的。

due to

due to 是"应归于……"或"由……引起（caused by）"的意思。后面要跟名词，与 because of 不同，但 due to 表达的也有因果关系的意思。如果不是很确定，应避免使用 due to，用 because of 或 caused by。

equipment

单数和复数都是 equipment，没有 equipments 的写法。

few，a few

few 是很少，强调没有多少，有否定的意思；a few 也是很少，但强调有一些，虽然不多，有肯定的意思。few 形容可数名词，little 形容不可数名词。little，a little 的用法与 few，a few 类似。

flammable，inflammable，nonflammable

flammable，inflammable 都是易燃的意思，nonflammable 是它们的反义词。

following

following 是形容词，表示"接着的，下述的"。

The behavior of the mouse was carefully monitored in the *following* days.

它也可以是动词 follow 的名词形式。

Following the flood, many wild lives found new habitant.

in contrast，on the contrary

都有"相反"之意，但 in contrary 是错误的。

induce，provoke

induce 引发；provoke 挑拨，科技写作中很少使用。

minimal，trivial

minimal 是最小的，trivial 是轻微的、不重要的。

percent，percentage，percentile

percent 跟在数字后面，以代替%，57 percent（57%）。percentage 是百分率，percentage 不能与数字一起用，可以说 small percentage or large percentage。Percentile 是一个统计学用语，表示在 100 个分组中事物出现的概率。

preceding，proceeding

preceding 前面的，proceeding 进程、事项。

present,represent

present 存在,represent 代表。

protect,preserve

protect 保护,preserve 保持不变。

proven,proved

proven 是形容词,被证明的,已证实的;proved 是过去式。

provided that,providing

provided that 假如,是连接词;providing 是现在分词。

remainder,remaining

都可以当"剩余"讲。

since

"自从……",表示时间;也可以用来表示因果关系,同 because。

subsequently,consequently

subsequently 表示时间顺序,consequently 表示逻辑推理结果。

such as,including

一般用逗号从主句分开,当 such as 后面排列的名词只有 1~2 个时,不影响句子的流畅,可以不用逗号分开。排列的事物和概念要相等。

Heavy metals *such as* lead are especially toxic to children.

The river is heavily polluted with heavy metals, *such as* lead, mercury, and cadmium.

such as,including 应紧跟它们要修饰的词组,不然就不明确。

In view of this inherent weakness, various surface modification techniques have been attempted by the present authors on austenitic stainless, such as laser cladding, laser alloying, with the goal of improving its cavitation erosion resistance and hence extending its service life in a cavitating environment.

宜改为:In view of this inherent weakness, we have investigated various surface modification techniques, such as laser cladding and laser alloying, to improve its resistance to cavitation erosion and extend its service life time in a cavitating environment.

The effects of the wood meals from different trees on the growth of HABs algae were assessed, including *Cunninghamia lanceolata* (Chinese fir), *Alnus cremastogyne*, *Pinus massoniana* Lamb, *Betula alnoides* and *Entandrophragma cylindricum* (Sapele).

宜改为：The effects of the wood meals from different trees, including *Cunninghamia lanceolata* (Chinese fir), *Alnus cremastogyne*, *Pinus massoniana* Lamb, *Betula alnoides* and *Entandrophragma cylindricum* (Sapele), on the growth of HABs algae were assessed.

symptom，syndrome

symptom 症状，syndrome 综合病症。

synergic，combined

synergic 协作的，more than combined。

than

用 than 来比较的两个对象要一致。

Acid A has a lower pK_a *than* acid B.

The water content in sample A is higher *than* that in sample B.

有些词代表最终状态，不能进行比较。这样的词有 full，absolute，complete，unique，extinct，permanent，universal 等。

that，which

that/which 经常用来引导修饰从句。如果修饰从句可以省略掉而不影响句子的完整性，也就是非限制性从句用 which，并用逗号把修饰从句与主句分开。

The recovered dogs, *which* were treated with antibiotics, are released.

如果修饰从句不可以省略掉，也就是限制性从句那就用 that，不需用逗号分开。

Dogs *that* were treated with antibiotics recovered.

限制性从句和非限制性从句在句子中的作用有很大区别。第一句话中，就是指"recovered dogs"，没有别的狗。第二句表示除了"dogs that were treated with antibiotics"，还有其他的狗。

toward，towards

toward，towards 的用法是一样的。美国用 toward，英国用 towards。

underlining，underlying

underlining 强调的，underlying（underlie）基本的。

upstream

upstream 是介词。You cannot say"in the upstream of"，"100 base pair upstream the coding sequence" is the right use。

use，using，utilize，employ

我们写作时经常会使用 using 这个词，但用 using 时经常会导致与主句的主语名词不一致，应避免使用。可用 with 或 by 取代。

<u>Using</u> the reductive reaction, reducing power of VC was examined.

宜改为：The reducing power of VC was examined *with* the reductive reaction.

utilize 是利用，有效使用的意思，与 use 不同，不能相互替代。

employ 雇用人，使从事于……

vary，change

vary 多样化，不同，种种；change 变化。

while

while 既可以用来做表示时间的连接词"当…的时候"，也可以用来表示转折和逻辑的连接词"然而、虽然、尽管"。为避免含义不清，用 when、and、but 或 although 取代更好。

4.2 用以描写研究课题的意义和重要性

用以描写研究课题的意义和重要性，常用于 Introduction 和 Discussion 的写作中。

associate with

S14 is a nuclear protein whose expression is closely *associated* with lipogenesis.

Angiogenesis is *associated* with diseases states such as cancer, diabetic retinopathy, rheumatoid arthritis and endometriosis.

These are *associated* with component models for the heat exchangers, receiver and radiator.

This pressure increase is *associated* with higher axial velocities, compared to the tangential velocity component, and therefore a lower swirl number.

attention

The chemical synthesis of genes and genomes has received considerable *attention* in the past and is becoming increasingly important in the exploitation of system biology.

The feasibility of increasing their operational range by attenuating resonance peaks and reducing vibration problems has received considerable *attention*.

Recently, the brittle fracture of PVC and other plastics has received increasing *attention*.

characterize

Alzheimer's disease (AD), the most common cause of dementia, is a complex neurological affection that is clinically *characterized* by the loss of memory and cognitive functions.

Atherosclerosis is a chronic inflammation *characterized* by endothelial dysfunction and lipid and microphage accumulation in the arterial wall.

composed of

CCMV is a well studied RNA virus with a shell *composed of* exactly 180 proteins.

Multiple agent systems are *composed of* a multitude of simple autonomous vehicles.

A hardware-in-the-loop simulation of a three-shaft gas turbine engine for ship propulsion *is composed of* computers, actual hardware, measuring instruments, interfaces between actual hardware and computers, and a network for communication, as well as the relevant software.

consist of

Turbogenerators used for power generation *consist of* several rotor stages that are rigidly coupled together to form a single rotor supported on many hydrodynamic bearings.

An immunoglobulin consists of two light chains and two heavy chains which are cross-linked together through Cys-Cys linkage.

evidence

There is growing *evidence* that phthalates in pregnant woman alter the reproductive organs of their male offspring.

Evidence for oxidative stress was detected among maples trees growing on acidic soil.

Such an increase in damping factor is *evidenced* by the reduction of vibration amplitudes.

There is growing *evidence* that designers and manufacturers of polyvinyl chloride (PVC)pipes are interested in the evaluation of fracture toughness under several operation conditions.

interest

One area that has continued to receive significant *interest* is the role of CBXC protein in the development of cancer.

There has been considerable *interest* in the study of pyrrolidine derivatives as inhibitors of protein kinase K.

We are *interested* in the rotor and foundation equation sets corresponding to synchronous motion.

For some considerations the unsteady behavior of the whole engine is of *interest*, i.e.the unsteady interactions between the individual engine components.

involve

This factor is *involved* in the growth of a variety of human tumors, including lung, breast and colon tumors.

Ghrelin, an gastrointestinal hormone, is *involved* in regulating food intake and body energy balance.

Specification of a turbocharger for a given engine *involves* matching the turbocharger performance characteristics with those of the piston engine.

The intake and exhaust ports are *involved* in the operation of the two-stroke and four-stroke cycle engines.

participate

Chaperone proteins *participate* in diverse cellular processes, including peptide translocation, stress response, and signal transduction.

Students were randomly selected to *participate* in this study.

When ignited, a fuel mixture of gasoline and air *participates* in the rapid combustion.

relate

Friction is intimately *related* to both adhesion and wear.

This increased rate of detection may be *related* to the fact that the patients examined were younger.

The four first eigenmodes of the compressor are overdamped and *related* to rigid body motions.

This frequency limitation is *related* to the adopted control strategy and not to the servo valve eigenfrequency.

role

GLP-1 plays an important *role* in the regulation of glucose metabolism.

Insulin is widely recognized to play an essential *role* in the regulation of glucose metabolism.

The simultaneous measurement of multiple species concentrations allows the calculation of local mixture fractions, which play an important *role* in ignition and flame propagation properties.

The acoustic response of the system surely plays a key *role* for the sustenance and amplification of a periodic instability.

4.3 用以描述某个领域的现状或作者的研究计划

用以描述某个领域的现状或作者的研究计划，常用于 Introduction 和 Discussion

的写作中。

characterize

In the present study, we try to *characterize* the role of pathogenic mtDNA mutations in the promotion of cancer.

Characterization and expression of two matrix metalloproteinase genes during sea urchin development is reported in this paper.

This sequence of events tends to counteract the initial situation, which is *characterized* by an intense inner recirculation zone and high mixture fractions only near the fuel nozzle exit.

These data *characterize* the relevant properties responsible for the formation of periodic instabilities.

demonstrate

Previous studies have *demonstrated* that glycemic control is important for the treatment of diabetic patients.

These findings *demonstrate* that pressure plays a critical role in both formation and reversibility of amorphous zirconium.

This *demonstrates* that we can study the transient characteristics of a gas turbine engine using this platform.

Although the mechanisms responsible for amplification and damping of combustion-driven acoustic oscillations are far from being fully understood, successful control of instabilities has been *demonstrated*.

discover

It is of great interest to *discover* a set of genes that can be used as tumor predictors.

The Big Bang theory is one of the most important *discoveries* of this century.

We have *discovered* that the tensile test behavior of PVC is affected considerably by the crosshead speed and orientation of the specimen.

It is of great interest to *discover* that the constant strain method only required two iterations to produce the earless target cup.

elucidate

Some of these compounds, especially X, may be useful tools in *elucidating* the properties of proteins.

Neuroimaging methods were used in this study to *elucidate* mechanisms of speech processing and reading in healthy and dyslexic populations.

It is very important to identify the basic parts of an engine and *elucidate* how they work together for the beginner.

Here we have *elucidated* how the two-stroke and four-stroke cycle engines operate.

establish

It has been *established* that GABA receptor played a role in memory storage.

The stereochemistry of Mitomycin was *established* by its chiral synthesis.

Establishing the dynamic characteristics of a gas turbine engine is an important task during its development.

Further investigation is needed to *establish* the viability of measuring the slope sufficiently accurately.

evaluate

We *evaluated* its contribution to the high background radioactivity in the region.

The aim of this study was to *evaluate* the current practice of peri and postoperative antithrombotic therapy in vascular surgery in Austria.

In theory, identification of the unbalance state, the configuration state, and the foundation is feasible, provided one can rely on the Reynolds equation for hydrodynamic lubrication to *evaluate* bearing forces.

In order to *evaluate* the heat fluxes between the fluid and the solid body of the compressor, some cutting planes have been positioned equidistantly.

examine

In this study, we have *examined* this highly unusual biosynthetic rearrangement in a series of expression experiments.

Examination of the antibody stained tissue showed a strong presence of MC5 receptor in the thalamus region of the brain.

In all cases *examined* the flow pattern tended to move at slightly less than rotor speed, with estimates of 90% to 97% of rotor speed being typical.

The effects of specimen orientations L-C, and C-L for specimen are *examined* at room temperature and at different crosshead speeds with the range.

explore

To further *explore* the effects of pressure, we performed an additional experiment at 14-17 GPa.

Here we *explore* the biogeographic consequences of presumed low O_2 levels during this period.

With the aim to identify which turbocharger is suitable for different types of engine duties, we *explore* the design-point and off-design-point performance of the overall thermodynamic cycle.

Here we *explore* the major differences between diesel and gasoline engines.

find

Cyclosporine was *found* to be cytotoxic to rat hepatocytes with an in vivo LD_{50} of 4.3 mg/kg.

We *find* a digital technique that can be used for art authentication.

Least-squares regression analysis can then be used to *find* the optimal foundation parameters.

The highest axial velocities are *found* between the inner and outer recirculation zones at radial positions between 5 and 15 mm.

identify

Ghrelin was *identified* as a critical hormone in the regulation of food intake of humans.

In the multi-race population of the US, many people *identify* themselves as more than one race.

Identification of mutations is a critical step toward understanding the biological functions of chaperone proteins.

The turbocharger compressor maps are used to *identify* minimum airflow, limit for continuous operation and minimum compressor efficiency.

investigate

The possible relationship between hepatitis C infection and liver cancer was *investigated*.

In this report, we *investigated* the effects of lipid modification of peptide Y on its stability in human plasma.

A three-shaft gas turbine engine for ship propulsion is taken as an object to establish the hardware-in-the-loop simulation platform, with which the fuel control strategy is *investigated*.

Theoretical turbocharger matching is useful to approach a range of turbocharger frames, but final testing is essential to *investigate* the effect of different turbochargers on the overall design-point and off-design-point piston-engine cycle performance.

monitor

Magnetic Ho^{3+} ordering was *monitored* by neutron powder diffraction and optical Faraday rotation.

Similar approaches could be applied to non-invasively *monitor* oxygenation in many low-protein body fluids.

The performance of this fuel-injected engine was measured using the engine's electronic control unit and its RAM *monitor*.

This analyzer is used to *monitor* and to record the frequency characteristic and the temporal variation of the fluctuating pressure.

reveal

Genetic experiments *revealed* that its biosynthesis involves an unprecedented oxidative rearrangement.

Recent advances have *revealed* the enormous complexity of even the simplest tribological process.

At this operating point, the detail with a finer temperature resolution clearly *reveals* the temperature difference between the fluid and the solid body of the compressor.

The phase-resolved measurements *revealed* significant variations of all measured quantities in the vicinity of the nozzle exit.

study

The metabolic control of body weight has been extensively *studied*.

We *studied* the synthesis of NH_3 by heterogeneous catalysis.

"Hardware-in-the-loop simulation" is an effective tool to *study* the dynamic process of a gas turbine engine.

A parametric *study* has been carried out for three-dimensional conjugate calculations with variations of the mass flow and the turbine inlet temperature.

4.4 用以描述需要解决的问题

用以描述需要解决的问题，常用于 Introduction 的写作中。

address

To *address* the issue of selectivity, a control sample was added.

The principles of the Health Belief Model were used to develop materials to *address* the needs, barriers, and motivators of this audience.

A graph was added to *address* the construction of the intake and exhaust ports.

aim

The *aim* of this research is to provide new potent compounds. To this end, we applied a design strategy in which two different drugs were combined into one molecule.

The *aim* of this study was to evaluate the activity of diazepines as thrombin inhibitors. In this report, we first present test data on a variety of genomic DNA with various genotypes.

although

Although glacier retreat is mentioned in almost all assessments on climate change, the number of systematic studies is quite small.

Although the angular velocity ratio is used for calculations involving just one pair of gears, it is more convenient when working with a gear train to use the reciprocal of the angular velocity ratio.

Although CIM encompasses many of the other advanced manufacturing technologies such as computer numerical control, computer-aided design/computer-aided manufacturing (CAD/CAM), robotics, and just-in-time delivery (JIT), it is more than a new technology or a new concept.

approach

Other *approaches* have focused on the pathways and kinetics of the capsid formation. But the origin of icosahedral symmetry in viruses has yet to be fully elucidated.

We *approached* the problem through an integrated methodology by using analytical and biochemical methods.

Three theoretical *approaches* have been introduced to investigate frictional forces in sheared systems.

Theoretical considerations of matching turbocharger pressure ratio and mass flow with engine mass flow and power permits designers to *approach* a series of potential turbochargers suitable for the engine.

describe

In this report, we *describe* an analytical method that can be used to measure the level of lead in human blood conveniently in a doctor's office.

As *described* in Materials and Method, RNA was prepared from cultured endothelial cells.

Negative values *describe* a heat flux in opposite direction, from the fluid to the solid body.

This function best *describes* the curve through the points of constant turbine inlet temperature.

despite

Despite these studies, little is known about the degradation of GLP1 in human plasma.

Despite the high temperature of the environment, these heat-shock proteins remained active.

Despite the high compression ratio, much of the heat energy is lost in waste exhaust gases, lost to friction, or lost in heating up the engine.

Despite the limitation of rotor displacements in the bearing gap, it is possible to reduce the rotor vibration by active forces acting on the bearings.

however

It is clear that adopting farming methods that enhance population density of wild plant and animal species on farm land is beneficial to biodiversity, provided that the change of to wildlife-friendly farming does not require a reduction in crop yield. *However*, it is frequently observed that the diversity value of farmland declining with increasing yield, suggesting that maintaining high wild life interest on farmland often conflict with high crop yields.

A number of glaciers have been studied by numerical modeling (3). *However*, the input data required for these methods is not available for most glaciers considered here.

Up to the present time it has been unnecessary to assign an algebraic sign to angular velocity ratio of a pair of gears. *However*, when gears are combined to give a gear train, it is important to consider the sign because it indicates direction of rotation. This is especially true in the analysis of planetary gear trains.

in contrast

In quantum dot devices, single electron charges are easily measured. Spin states in quantum dots, however, have only been studied by measuring the averaging signal from a large ensemble of electron spins. *In contrast*, the experiment presented here aims at a single-shot measurement of the spin orientation of a particular electron.

Current switches, *in contrast*, tend to control a weak beam with a strong one.

The emissivity obtained with the widebandpass filters has a tendency to increase with increase in temperature. *In contrast*, the emissivity obtained with the narrowbandpass filters has a tendency to decrease with temperature.

This mechanism of heat transfer is often utilized to heat other materials because of the high emergent heat flux. *In contrast*, when heat is released as a radiative heat flux, the temperature of the surface of the material is reduced, thus preventing thermal degradation or oxidation of the material.

present

This article *presents* research data on the advantages and limitations of the methods that are available to measure environmental tobacco smoke exposure.

In this report, we first *present* test data on a variety of genomic DNA with various genotypes.

In the paper a feasibility study of inverted Brayton cycle (IBC)engines, for the repowering of existing gas turbines, is *presented*.

The turbine performance maps are *presented* as specific power versus mass flow rate.

prompt

These findings *prompted* us to investigate the relationship between mitochondrial dysfunction and carcinogenesis.

The application *prompted* us to predict the transient response of a small solar-powered regenerative gas-turbine engine with centrifugal impellers.

This finding *prompted* us to discuss the position measurement requirements, mechanical design, fabrication, and alignment issues encountered for both sets of Beam Position Monitors.

purpose

For optimal balancing and diagnostic *purposes* it is important to be able to correctly predict the system vibration behavior over the operating speed range.

The *purpose* of this study was therefore to compile an extensive experimental database on a combustor with practical relevance.

remain

The production of bulky pure glass materials in pure metals *remains* a long-standing scientific curiosity and technical interests.

The mechanism of Peptide Y's biological functions *remains* unclear.

Despite its practical and fundamental interest, the origin of such a system *remains* an unsolved problem.

Under lean conditions, OH concentrations *remain* high as long as the temperature is high.

report

A series of compounds, including B, were subsequently *reported* by scientist at MIT.

As *reported* by Mason et al. that Leptin played a key role in human obesity.

This paper *reports* a fuel control simulation implementation of hardware-in-the-loop system for gas turbine.

The observation of a variation of heat release as a consequence of fluctuating mixture fraction has also been *reported* by others.

represent

To our knowledge, it *represents* the first reported porphyrin conjugate in which the peptide is directly linked to the porphyrin.

The solid line in Figure 4 *represents* the changes of fluorescence intensity over time.

This is a major task owing to the difficulty in properly *representing* the many joints and ground effects.

It is assumed that one has an adequate model of the rotor, which is axially symmetric and can be adequately *represented* by system matrices.

unfortunately

The contribution of glacier to sea-level has been estimated. *Unfortunately*, such observation started only in the first half of the 20th century and do not provide information about the transition from the Little Ice Age to the current climate state.

Unfortunately, because the nonlinear optical interacting strength of most materials is so small, achieving single-photo switch is difficult.

Unfortunately, inherent in the use of statistical sampling procedures is acknowledgement of the risk that some defective parts will slip through.

4.5 有关过去和现在时间的词汇

有关过去和现在时间的词汇，常用于 Introduction 和 Discussion 的写作中。

current

There is *currently* no diagnostic method for early detection of Autism.

Current scientific evidence indicates that dietary fat plays a role in weight loss and maintenance.

The *current* work discusses means to utilize low-grade small-scale energy in vehicle exhaust gases, to reduce the vehicle's fuel consumption and to make it run more environmental friendly.

present

The *present* study aims to explore the barriers faced by primary care physicians in the management of patients with chronic hepatitis B infection.

previous

The measurement of nicotine and cotinine levels in the appendages of the skin (hair and nails) reflects exposure to tobacco over the *previous* three months and could become a better reference marker in epidemiological and toxicological studies.

Previous studies demonstrated a linear relationship between the longitudinal relaxation rate and oxygen content, which permits quantification of the CSF oxygen partial pressure.

The *previous* work in topology design of compliant mechanism coupling structures for piezoelectric actuators has focused mainly on quasi-static applications.

The steady state model may be extended to a dynamic one by assuming a quasi-steady state approximation, as was done in the *previous* models.

recent

The *recent* discussion on the discrepancy between surface temperature and satellite measurement demonstrates the importance of high-elevation climate.

It was *recently* predicted that even in the perfect conducting limit, the same metal can support localized waves.

Examples of *recent* work include the use of topology optimization to design compliant mechanisms is extracted and applied to linkage number synthesis.

to date

To date, however, early education therapy is the only clinical effective approach.

It is therefore necessary to discover pharmacological compounds that are effective in treating this disorder.

This sequence appears to be the longest synthetic DNA reported *to date*.

Research *to date* has shown that using diesel/water emulsion as an alternative fuel to standard diesel fuel can significantly reduce the emission levels of both pollutants if the penalty in reduced power can be tolerated.

4.6 用于举例的词汇

用于举例的词汇，经常用于 Introduction，Results 和 Discussion 的写作中。

case

Extreme caution should be taken when handling strong acids. *In case of* contact exposure, wash the contact area immediately with running water.

For the specific *case* of carburetors, this model captures phenomena reported in the literature.

like

The main reason for the application of heavy ions *like* carbon in radiotherapy is the enhanced relative biological effectiveness.

Arabic countries *like* Jordan and Syria are facing the rise of Islam.

Especially for automobile air conditioning in EU, CO_2 has more priority than other natural refrigerants *like* NH_3 and hydrocarbons.

The choice of this mix allows a reduction of reboiler heat duty, as the MDEA requires less heat for regeneration than other amines, *like* MEA and DEA.

especially

Taken together, these data indicate that agriculture is the major current and likely future threat to bird species, *especially* in developing world.

Teenagers, 12-19, *especially* girls, are the most victimized segment of the population in the United States by internet crimes.

The higher fuel rates, *especially* in the BIOM (a) case, suggest a mechanical and thermal output increase which must be confirmed by a more accurate analysis based on the component matching.

for example

For example, Schleiser et al. considered a theory of canonical-dissipative systems.

Although these temperature differences are small, *for example*, the difference in the maximum temperature is only 27 K, they can cause substantial discrepancies in the predicted NO magnitudes.

for instance

Changes in agriculture are more pronounced in the developing world. *For instance*, the total area of cropland in the developing world has increased by over 20% since 1961, whereas developed world cropland has shrunk.

For instance, the ith filter uses the sensor subset yi that excludes the ith sensor, where i is an integer from 1 to m.

include

High sodium intake predicted the risk of type 2 diabetes, independently of other risk factors *including* physical inactivity, obesity and hypertension.

Many races *including* Black and Hispanic are considered minority in US.

Other alkali containing particulates *include* silicates and aluminosilicates.

The comprehensive database *included* mean and fluctuating velocity components, mean temperature, radiation heat flux, as well as species and pollutant concentrations.

in particular

However, proton and ion therapy have their own clinical and technical uniqueness: *in particular*, they require definition of new clinical, radiobiological and technical concepts, and strict quality assurance programs for efficient and safe clinical application.

Marriage after a cancer diagnosis has long been neglected in the clinical and research settings in many countries. *In particular*, research in Asia on marriage and cancer among women is extremely scarce.

In particular, the transmittance is observed to decrease in the upper and lower edges of the two images.

In particular, some maps of the COE and break-even carbon tax (BECT) behavior have been constructed to test the importance of the market uncertainty on the economic results obtained.

particularly

Another rapidly growing area of tribology is in biosystems, and *particularly* the lubrication mechanisms in joints.

The climatic information contained in records of glacier geometry, *particularly* glacier length, has only partly been exploited.

Poor agreement of the EDS-finite-rate model is also observed, *particularly* at two upstream sections.

At downstream sections, the trend and magnitude are fairly correlated with the measurements, *particularly* for the EDS model results.

specifically

The variability and sensitivity were within range of the accuracy of our procedure; *specifically*, less than 1.2% variation and 1 ng/mL sensitivity were found for this assay.

We made efforts to search for variable stars. *Specifically*, 28 new variable stars were found in the one degree field of NGC 7789.

The method and algorithm, which are *specifically* designed for brush seal applications, have been coded into a new computer program, which can be run stand-alone or as a "plug-in" for a CFD code.

Specifically, fine particles are a concern because they are less likely to be captured by electrostatic precipitators, cyclones, or baghouses than their larger counterparts.

such as

GM varieties are primarily used for industrial crops, *such as* cotton, and for animal feed crops.

Countries, *such as* China and India, are playing an increasing role in the manufacturing of pharmaceutical ingredients.

In the basic cycle analysis, several assumptions are made, *such as* compression and expansion efficiencies and so on.

Syngas is the raw material for several chemical syntheses, *such as* methanol and ammonia.

4.7 用以描述实验部分的词汇

用以描述实验部分的词汇，常用于 Experiment 和 Results 的写作中。

according to

Cells were transfected using SuperFect (Qiagen, Valencia, CA) *according to* manufacturer's instructions.

5-Nitropurine was prepared *according to* literature procedure as colorless crystals, mp 178-180 ℃.

The G matrix was calculated *according to* Eq. (3).

achieve

This mild condition *achieved* the transformation of carboxylic group while leaving the CBZ group intact.

Although growth in global food production exceeded population growth between 1961 and 1999, this was *achieved* through a 12% increase in the area of cropland and a 10% increase in the area of permanent pasture.

The fine particle number concentrations for cofiring were much higher than those *achieved* with dedicated coal combustion.

To *achieve* stable dilution ratios, the rates of dilution N_2 and exhaust flow were maintained by critical flow orifices.

acquire

We *acquired* PE spectra of a number of water clusters.

To *acquire* new information on copper metabolism, we utilized cuprizone, a very sensitive and selective copper-chelating agent, as a relevant chemical model in mice.

Data were *acquired* at 40 kHz with the low-pass filter set at 20 kHz (Nyquist) frequency to prevent aliasing.

The a*cquired* data were subsequently transferred to a computer for post-processing.

afford

The Boc group was then removed by treatment with HCl in dioxane, *affording* the sulfonamide S.

Replacing the phenyl group with lipophilic residues such as 3,4-dimethylphenyl also *afforded* high affinity ligands, as did bicyclic aromatic group.

allow

Cells were seeded in 96-well plates and *allowed* to grow for 24 hours.

This wash step *allowed* the detection to be performed at room temperature.

Thus the matching analysis *allows* a more realistic estimation of actual variations which can be expected in plant performances and emissions.

This *allowed* us to determine the temperature from each image on the basis of two-color 2D thermometry.

assess

The migration effects of PRL3 were *assessed* by infecting 5000 human endothelial cells for 2 days with a multiplicity of infection of 300.

The authors analyzed data from 2 Phase Ⅲ trials to *assess* the effect of the chemotherapeutic agent.

In order to *assess* dynamic flow conditions, the model was extended by solving instantaneous one-dimensional Navier-Stokes equations in single-phase pipes.

carry out

LC-MS analysis was *carried out* using API4000 mass spectrometer.

Patient treatments were *carried out* by radiation oncologists of the University of Miami.

Due to a lack of confidence in predicting the velocity and temperature distributions in combustors, it is difficult to *carry out* a reliable evaluation of NO_x models applicable to practical combustors.

However, for bevel planets, in order to *carry out* the dot product to compute power, it is necessary to know the angle between the absolute angular velocity vector of a planet and its applied torque.

collect

One hundred and eighty primary ovarian tumor samples were *collected* in the hospitals of New England. The collection and use of tissues for this study were approved by appropriate institutional ethics committee.

The major species measured were CO, CO_2, H_2, and C_3H_8. NO_x and NO were *collected* through the same probe but analyzed using a Scintrix NO_x analyzer.

conduct

The field trial was *conducted* by farmers in one village in Shandong province.

Three classes were randomly selected to participate in a 6-week prevention program *conducted* by the trained teachers.

Based on this knowledge, the isolation of faults is *conducted*.

The experiments were *conducted* in a two-stage axial turbine.

do

Reverse transcription was *done* using Superscript II (Invitrogen) and 5 μg of total RNA from cultured cells.

A similar equation could be written for the air system, but it was not *done* considering that the inertia of the fuel is three orders of magnitude larger than the inertia of air.

expose

We *exposed* the thin film to a series of gas at T=500 K.

It is well known that children who were *exposed* to lead showed impaired learning ability.

After further rotation, the channels meet the second-inlet port through which the high pressure low-temperature water comes in and is *exposed* to the low-pressure high-temperature superheated vapor in region.

from

Antibodies were *from* Santa Cruz Biotechnology, Inc. (Santa Cruz, CA).

furnish

This new bromination procedure selectively *furnished* the corresponding monobromination products.

All blood providers are required to *furnish* identifying information on themselves which would then be made available to their doctors.

generate

The data were used to *generate* a three dimensional images of the tissues.

Geological evidence demonstrates that liquefaction due to strong ground shaking, similar in scale to that *generated* by the New Madrid earthquakes, has occurred at least three and possibly four times in the past 2,000 years.

The first step is to generate a graph of the mechanism. From this graph, an adjacency matrix is *generated*.

A pump was used to supply mineral spirits to the float valve of the carburetor at approximately the same pressure that would be *generated* by an elevated fuel tank.

keep

The temperature was *kept* constant at 45 ℃ for one hour.

Substrate concentrations were *kept* between 100 to 500 ng/mL for the assay.

The amount of fuel is reduced to *keep* the engine operating at "steady state".

This effect is eliminated with the unsteady model, since fuel has enough inertia to *keep* flowing even at very low quantities.

obtain

Electron tunneling parameters have been *obtained* from analysis of the decay kinetics.

Because of the uncertainties of measurement, we can not *obtain* adequate data for each solvent.

The local values of temperature and species concentrations of O_2 and N_2 are *obtained* from the flow-field simulation.

The numerical results were *obtained* with the thermal and prompt NO mechanisms, the partial-equilibrium O and OH approach, as well as the turbulence effect.

perform

Analytical thin-layer chromatography(TLC) was *performed* on Merck precoated silica gel 60 F254 plates and visualized by ultraviolet (UV) illumination (254 nm) using a UV GL-58 mineral-light lamp.

Optical measurement was *performed* using previously described transmission set-ups.

Some field tests for combined microturbine and desiccant airconditioning units are *performed* in Japan.

Computations have been *performed* for a hypothetical single-cylinder version of a representative, heavy-duty, truck diesel engine.

prepare

A polyclonal antibody against BCD1 was *prepared* as described previously.

In the current study, we compared different tissue *preparation* techniques prior to analysis with respect to their suitability for protein retrieval.

The site development costs, which consist of material, land, freights, and manpower labor, concern the activities that are necessary to *prepare* the site where the power plant will be located.

produce

Compound Y *produced* a dose dependent response.

Most current commercial flow cytometers employ analog circuitry to provide feature values describing the pulse waveforms *produced* from suspended cells and particles.

The adoption of different fuels *produces* changes in the mass flow rate through the turbine.

The selected motors are direct current motors with a gear ratio of 18.5:1 and are able to *produce* torques of 4.5 N • m at 155 rpm.

proceed

Catalytic hydroxylation of vinyl group *proceeds* in high yield to give the final product.

The reactions catalyzed by esterases are proposed to *proceed* in a two-step process.

provide

We *provide* experimental evidence of the resonance of a surface mode propagating across the surface with holes.

The proper predictions of velocity and temperature fields *provide* a reliable base to evaluate the NO emission models.

The contour plot *provides* a whole picture of comparison between the numerical and experimental results.

Tumor DNA from 70 primary breast cancer patients was *provided* by Dr. Nick Nelson (University of Florida, Miami, FL).

purchase

Unless stated otherwise, all chemicals were *purchased* from Sigma Chemical Company (St. Louis, MO).

The rates for *purchased* electricity and natural gas adopted in this study are given in Table 2.

record

Nuclear magnetic resonance (NMR) spectra were *recorded* on a 200 MHz Bruker spectrometer. ^{1}H NMR spectra were *recorded* in parts per million (ppm) in DMSO-d_6 or $CDCl_3$ with TMS as the internal standard.

Our initial set of measurements was *recorded* at a series of fixed values of temperature.

Repeated time *records*, each spanning one complete rotor disk revolution, were acquired for 50 rotor revolutions.

A second microphone probe in one of the rods holding the windows allowed *recording* the acoustic signal in the combustor.

subject to

Cells were rinsed in PBS, lysed in assay buffer, and *subjected to* Western blotting as described previously.

The device is *subjected to* a magnetic field of 10 T.

The boost pressure p_2 is first estimated for the target power output *subject* to thermal and mechanical stresses.

treat

One gram of nitrophenol was *treated* with 0.3 mL of benzaldehyde in 100 mL methanol at 50 ℃ for 5 hours.

It must be pointed out that the cycle analysis neglects the energy requirements for biomass or solid-waste *treatment*.

undertake

We have thus *undertaken* a systematic study of the effects of collisions of the inert gas atoms and small molecules.

We *undertook* this study to understand the community members' knowledge, attitudes, and practices related to AIDS prevention.

yield

Plotting all records *yields* a curve for the change of water levels.

The reaction of diphenylacetylene and biphenylene at 80 ℃ in the presence of a catalyst *yields* diphenylphenanthrene quantitatively.

Since CH distributions are a qualitative measure for local heat release rates, integration of the CH LIF intensities over an image *yields* a relative total average heat release rate at each phase angle.

Thus the technique just described *yields* three equations from Eq.(45), two equations from Eq.(47) , two equations from Eq.(48) for links 2 and 3, and one equation from Eq.(49) for the input link 1.

4.8　用于描写先后顺序的词汇

用于描写先后顺序的词汇，常用于 Experiment 和 Results 的写作中。

follow

The compound was prepared via a literature procedure by reduction of benzidine, *followed by* hydrolysis with NaOH.

Azathioprine (AZA) is a thiopurine prodrug commonly used in triple-immunosuppressive therapy *following* liver transplantation.

It can be noted that both airflow and fuel flow closely *follow* the pressure variation.

Kochev[11] constrains the mechanism to *follow* a predefined trajectory chosen in such a way that no force or moments are generated at the base.

immediate

Subjects ate a typical "western diet" (60% carbohydrate, 30% fat, 10% protein) for 2 days, *immediately* followed by 7 days of an high fat diet (5% carbohydrate, 60% fat, 35% protein).

The reaction was removed from the water bath and *immediately* quenched with ethanol.

instant

Sonography was performed *instantly* after brain removal.

These problems of vomiting and aspiration that are associated with anesthesia require *instant* recognition and a rapid, appropriate response by all anesthetists.

in the course of / during the course of

In the course of our study, we measured the DNA binding activities of these compounds by fluorescence assay.

During the course of these investigations, other (possibly interesting) observations were made.

simultaneous

The best is to *simultaneously* deliver food production needs and meet conservation goals.

We report a method for *simultaneous* determination of six HIV protease inhibitors in human plasma by high-performance liquid chromatography.

Internal combustion engine designs are continuously moving towards increased engine power, reduced engine size and improved fuel economy, *simultaneously*.

A linear perturbation approach is used for solving Eq. (8) due to its computational efficiency in *simultaneously* calculating the oil film nominal pressure as well as stiffness and damping matrices.

subsequent

The compound was prepared via a literature procedure by reduction of ketones and *subsequent* hydrolysis with NaOH.

The water level reached the maximum height and *subsequently* receded.

Subsequently, the resistance torque, which is the result of external loading imposed on the engine by the vehicle or the dynamometer, is subtracted from the brake torque and the net value is passed on to the engine dynamics sub-model.

then

The microplate was shaken for 5 minutes at room temperature and *then* incubated at 4 ℃ for 24 hours.

When air flows through the air-bleed system, it passes through an air-bleed orifice at the entrance of the venturi, *then* through a series of small passages, and ends up at the top of the fuel well.

The general approach used here to create and solve a carburetor model is to solve for airflow first, which *then* sets the boundary conditions for the fuel flow network.

4.9 描写实验结果的词汇

描写实验结果的词汇，常用于 Results 和 Discussion 的写作中。

affect

Even a low dose of the drug can *affect* the ability to function properly.

No technology promises to *affect* our world more profoundly than the rapid sweep of digital technology.

This characterization is important for the overall HCCI dynamics because the thermodynamic state (pressure, temperature) and concentration (oxygen and inert gas) of the exhausted mass flow *affect* the next combustion event.

Results will be presented to illustrate how the above features *affect* device performance for the frequency range of 4 kHz to 21 kHz.

alter

The extent of natural habitats on agriculturally usable land has been *altered* by clearance for cropland and pasture.

Incorporation of histone variants into nucleosomes provides another mechanism for *altering* chromatin structure.

Pressure variations in the plenum or the combustor *alter* the inlet air and fuel spray characteristics.

This can drastically *alter* the life of the blade and possibly instigate early fatigue failure.

analyze

The major metabolite was *analyzed* by HPLC on an Agilent HP1100 (Agilent Technologies, Palo Alto, CA) series instrument with a UV detector (229 nm).

In the present study we *analyzed* the potential role of changes in progesterone levels in the initiation of child-birth.

An experimental study is performed, in order to acquire and *analyze* test data.

Data covering an extensive range of the compressor performance map have been collected and *analyzed*.

confirm

The results *confirm* the qualitative conclusion that tolune mediates long-range electron tunneling more efficiently than tetrohydrofuran.

Loss of forest, especially in tropical regions, was *confirmed* by satellite image.

We *confirm* the recently detected linear correlation between radio and optical core emission in FR I galaxies and show that both core emissions also correlate with central Hα+[N II] emission.

consider

Whereas circulating Protein X was detected in 70% of the patient samples, only 38% were *considered* increased compared with the non-patient group.

This algebra is generated by the matrix elements, *considered* as functions on G.

By *considering* the algebra of polynomial functions, we get an easy comultiplication.

As a test case, we *consider* the nearfield plume flow for a Star-27 solid rocket motor exhausting into a vacuum.

detect

Particularly for applications where the information carrying sensor nuclei are in a very dilute concentration, such as medical imaging (e.g., lung MRI), NMR of porous void spaces, and biomolecular binding events in solution, the *detection* sensitivity is poor for the NMR spins of interest.

The specificity of direct sequencing and of the p53 GeneChip assay at *detecting* p53 mutations were 100% and 98%, respectively.

This paper deals with the transient thermal signal around an engine cylinder in order to propose a new and nonintrusive method of knock *detection*.

determine

To investigate the relation ship between PW expressions and proliferate activity of these cells, we *determined* the rate of proliferation by thymidine incorporation assay (Fig.1B).

It is clear that the hydrogen bond network and its fluctuations and rearrangement dynamics *determine* the properties of the liquid water.

Numerical simulations for the three-dimensional flow field and particle trajectories through a low-pressure gas turbine are employed to *determine* the particle impact conditions with stator vanes and rotor blades using experimentally based particle restitution models.

In terms of both the frequency response of the transducers, and hardware capabilities which *determine* the resolution and dynamic range.

develop

We *developed* a finite element model that simulates coseismic slip associated with earthquake.

As part of a project to *develop* an agonist of insulin receptor, we synthesized a series of peptides.

Actually this work tried to *develop* an affordable process to integrate cheap sol-gel deposition process with silicon technology.

distinguish

Carbon-14 dating clearly *distinguishes* objects found in the tomb and samples nearby.

Our results supports the notion that mammals use a combination of roughly 1,000 types of olfactory receptors expressed on olfactory neurons to *distinguish* about 10,000 different odors.

The OCR engine uses features that are based on the Fourier descriptor to *distinguish* characters.

The Bayesian odds ratio cannot *distinguish* between the two model distribution functions for the two smaller burst samples with spectroscopic redshifts.

exert

Although both compounds contain the glutamic acid functional group, there is little evidence to suggest compound 1 and 2 *exert* their pharmacological activity through direct interactions with glutamic receptors.

Our results indicate that actin *exerts* retractile or propulsive forces depending on the local membrane curvature and that the membrane is strongly bound to the actin gel.

These additional terms should *exert* a significant influence in some rarefied gas phenomena occurring in micro channel flows or in atmospheric re-entries of spatial engines.

Substantial decrease in stroke and thrust of the actuators needed to move the table or *exert* forces on the structural masses.

exhibit

Single-walled carbon nanotubes with diameters in the range from 1 to 2 nm *exhibit* unique electronic behavior.

We also found that R *exhibited* the same time behavior as S and they are linearly related.

In the proposed study, a radically different concept is presented whereby periodic mounts are considered because these mounts *exhibit* unique dynamic characteristics that make them act as mechanical filters for wave propagation.

The possibility of generalizing the validity of observations is supported, by presenting results from testing a second turbocharger, which is shown to *exhibit* similar behavior.

illustrate

Data shown in Fig 3 clearly *illustrated* the high specificity of the assay.

As *illustrated* in Fig 1, the fluorescence intensity is nearly linear to the substrate concentration.

Experimental data and model predictions *illustrate* that a significant reduction in bearing friction can be achieved.

interact

We conclude that glucose does not directly *interact* with the GL-1 receptor in an in vitro expression system.

The exhaustive research into prostate cancer to date has demonstrated a complex *interaction* of multiple genes and environmental factors.

As these dilution jets *interact* with the mainstream flow, kidney-shaped thermal fields result due to counter-rotating vortices that develop.

lead to

In summary, our exploration of the SAR of indole derivatives *leads to* the finding that a sulfonamide group is essential for activity.

Consumption of high carbohydrate diets *leads to* increased mRNA expression of enzymes that are involved in converting glucose to fatty acids.

Also, it *leads to* development of specific laser propulsion engines running on auxiliary fuel with improved LPE thrust characteristics.

Simulation results show that rotation may *lead to* slightly higher particle temperatures near the central axis, but for the case considered the effects of particle rotation are generally found to be negligible.

manipulate

The ability to control and *manipulate* frictional forces is extremely important for many applications.

Additional characterization of these particulate structures provided a potential mechanism to *manipulate* the adhesive forces with which these particles are bound to the surface.

The use of a robot to *manipulate* parts under inspection, a high-frequency pulse generator for inductive heating and enhanced algorithms enabled a demonstrator to be set up for the fully automated crack inspection of engine compressor blades.

User/Analyst will be able to *manipulate* models directly through a variety of tools.

mediate

Caspase 8 is a key enzyme in death receptor-*mediated* apoptosis.

Here we show that the CD28 antigen *mediated* specific inter-cellular adhesion with human lymphoblastoid and leukemic B-cell lines.

A striking aspect of the recently proposed split supersymmetry is the existence of heavy gluinos which are metastable because of the very heavy squarks which *mediate* their decay.

modify

The peptide was *modified* by a cysteine-specific fluorescein derivative.

Elevated plasma LDL levels, followed by oxidative *modification* in the arterial wall, could sufficiently account for the initiation well-defined lesion in atherogenesis.

Readers interested in a detailed understanding of the techniques used to *modify* the PART5 model are referred to the final project report to TNRCC (Eastern Research Group 2000).

The system uses flow injection to *modify* temperature and pressure inlet distortion.

modulate

A growing body of literature suggests that the melanocortin (MC) system *modulates* neurobiological responses to drugs of abuse.

Oligonucleotide DNA arrays were employed to identify genes that were differentially *modulated* by the overexpression of TIMP-3.

One practical method for achieving homogeneous charge compression ignition (HCCI) in internal combustion engines is to *modulate* the valves to trap or reinduct exhaust gases.

A 0-450 Hertz bandwidth, voice coil actuated, proportional sleeve valve is designed to *modulate* air mass flow by controlling the throat area of a choked flow.

observe

In the absence of water, we *observed* a striking increase in the stability of Prizil at ambient temperature.

A significant reduction of proliferate activity was *observed* after antibody treatment.

Aerodynamic measurements were acquired on a modern single-stage, transonic, high-pressure turbine to *observe* the effects of low-pressure turbine vane clocking on overall turbine performance.

We do not *observe* evidence of absorption lines in the *XMM-Newton* or reprocessed *Chandra* data.

present

The cells treated with vehicle alone did not *present* any morphological change (Fig.2B, E and G), suggesting there is no ligand-independent effect.

In a classical view, MHC-I molecules *present* peptides from intracellular source proteins, whereas MHC-II molecules *present* antigenic peptides from exogenous and membrane proteins. We demonstrated that peptide *presentation* is altered considerably upon induction of autophagy.

result in

Substitution of position A with methyl group *resulted in* similarly increased receptor activity.

The DNA damage *results in* cell-cycle arrest in either G_1 or G_2 phase. The DNA damage also *results in* slowing of DNA replication and cell-cycle progression when DNA damage occurs during S phase.

This observation is contrary to that obtained with normal temperature combustion air wherein diffusion flames *result in* higher NO_x emission levels.

The results indicate large penetration depths for the high momentum dilution jets, which *result in* a highly turbulent flow field.

retain

Our aim was to establish primary adult human RPE cell cultures that *retain* their epithelial morphology in vitro.

The purified enzyme when kept at 45 ℃ and 50 ℃ for 40 min *retained* 92% and 85% protease activity, respectively.

The current best theoretical stellar models, however, do not *retain* enough angular momentum in the core of the star to make a centrifugally supported disk.

Raytheon Infrared Operations (RIO) programs have produced a variety of products that are economically viable for the commercial market and *retain* very high performance.

show

As *shown* in Figure 3, the output in March is the highest in the whole year.

Plasma hormone concentration *showed* a marked rhythm peaking at 12 hours.

The approach proposed in this paper has been *shown* to be a powerful tool than manual numerical simulation procedure.

用来描述降低、减少、抑制、阻止的词汇。

decrease

It *decreased* the relative amplitude by 80% and 37%, respectively, in mice subject to chronic sleeplessness as compared with the control.

Around 1800, mean glacier length was *decreasing* and this *decrease* accelerated gradually.

Upon further advance of the injection timing, the efficiency *decreased* to about 40% at 60 deg BTDC.

Further, the peak heat release rates were also *decreased* as injection timing was retarded from 35 deg to 15 deg BTDC.

hamper

Application of these assays has been *hampered* by lack of specificity.

Proposed federal cuts in some areas of drug enforcement may *hamper* the ability of authorities to seize methamphetamine.

inhibit

After 48-hour treatment with Tomycin, CHO cells did not show significant reduction of invasive capacity compared with ethanol-treated cells. However, after treatment for 72 hours, the invasive activity of CHO cells was markedly *inhibited*.

The mice injected with the drug *exhibited* a severely altered glucose profile.

The presence of a significant cavitation zone can *inhibit* vorticity transport causing nearly all the fluid to be ejected through a crescent-shaped sector of the orifice exit plane.

Energy deposition by decaying neutrons may *inhibit* spherical accretion and drive a wind.

prevent

Nitrogen gas protection of the reaction *prevented* the loss of material from air oxidation.

Taking folic acid before and in the first weeks of pregnancy can help *prevent* certain serious and common birth defects.

The X-rays from the central object are significantly attenuated by the disk atmosphere so they cannot *prevent* the local disk radiation from pushing matter away from the disk.

reduce

We *reduced* the amount of energy applied for fragmentation to achieve a better sensitivity.

Drugs that can either *reduce* calorie intake or increase energy consumption are actively sought to control obesity.

One may *reduce* these problems by scanning entire lines at once, rather than points.

Catalytic converters are widely used to *reduce* the amounts of nitrogen oxides, carbon monoxide and unburned hydrocarbons in automotive emissions.

suppress

Injection of 2 mg/kg test compound X *suppressed* the activity of rat 1 and rat 2 by 30% and 43%, respectively.

People can learn to *suppress* pain when they are shown the activity of a pain-control region of their brain, a small new study suggests.

This paper describes the model-based design and the experimental validation of a control system which *suppresses* the bouncing behavior of Compressed Natural Gas (CNG) fuel injectors.

The results of the experiments indicate that both the adaptive controller and the gain-scheduled controller effectively *suppress* the noise of engine exhaust systems.

用来描述提高、增加、激活的词汇。

elevate

The expression of DNA K protein was *elevated* slightly after an increase of temperature from 25 ℃ to 30 ℃.

Inhibitors of the anandamide-deactivating enzyme, fatty-acid amide hydrolase which selectively *elevate* anandamide concentrations, exert similar effects.

A pump was used to supply mineral spirits to the float valve of the carburetor at approximately the same pressure that would be generated by an *elevated* fuel tank.

The mismatch between the modeling parameters and the actual hardware parameters explains why the experimental values for the shaped cam are slightly *elevated* about the theoretical predictions.

enhance

The electric conductivity was largely *enhanced* just after photo-excitation.

Electron spectroscopy was used to study the *enhancement* of absolute anion concentrations at the interface of salt solutions.

We also present an approach for selecting the form of a customer (choice) utility function of demand analysis to *enhance* the predictive accuracy.

increase

Loss of each additional carbon *increases* the tunneling rate by a factor of 2.4.

Addition of salt to water causes an *increase* of surface tension.

Surge has been successfully controlled in small centrifugal compressors yielding a 20%-50% *increase* in operating range.

A thermal protection system was sized to prevent pressure vessel aerothermal failure and was found to *increase* the mass of the NTR by approximately 15 percent.

raise

Raised cortico-steroid levels have been implicated as a possible causal mechanism for the neuro-psychological impairments in the early stages of dieting to lose weight.

With the identification of stem cell plasticity several years ago, multiple reports *raised* hopes that tissue repair by stem cell transplantation could be within reach in the near future.

New photoinjector technology should *raise* the output power to one megawatt, the design value for the laser.

Models of low-density gas in the narrow-line region show that relativistic particles can *raise* the temperature in the O_3 zone.

stimulate

DT56a, a natural compound for the treatment of menopausal symptoms and osteoporosis, *stimulates* bone formation in female rat.

UK scientists have found that videogames *stimulate* learning.

The new Onyx2 connection mesh architecture made it possible to develop a more economical system while maintaining the fidelity needed to *stimulate* actual sensors.

4.10 用于比较语句的词汇

用于比较语句的词汇，常用于 Results 和 Discussion 的写作中。

agree

In *agreement* with previous studies, lower crustal flow models lead to poor fits.

The results *agree* exceptionally well with those determined using a finite element method model.

The experimental results obtained on two generators with AC as well as DC loads closely *agree* with the values predetermined using the GA approach.

The dispersion curves *agree* with experimental frequency dependences of the excitation efficiency for various modes in the plates.

compare

The expression level was increased by 56% *compared with* the control.

In *comparison* with the best power-law models, Newtonian models predict the GPS observations poorly.

The mutant parasite retained its host cell invasion capacity at levels *comparable* to the wild-type parasite.

Fiber-optic sensor performance is shown to *compare* very favorably with co-located thermocouples where such co-location was feasible.

consistent with

The influence of the acidity on their color intensities are *consistent with* what was reported by Suggs.

The percentage of cells expressing B22C was *consistent with* earlier founding.

The suggestion that wake migration is the dominate mechanism in generating the clocking effect is also *consistent with* anecdotal evidence that fully cooled engine rigs do not see a great deal of clocking effect.

equivalent

Equivalent CO_2 concentration is a measure of radiative forcing expressed in terms of the concentration of CO_2 that would produce an amount of forcing *equivalent* to the total forcing from all gases and aerosols combined. We define the *equivalent* CO_2 level to be equal to the true CO_2 level in 2000.

After the addition of 1 mole-*equivalent* of Cu(Ⅱ), further additions of Cu(Ⅱ) had no effect on the CD spectrum.

In this paper, the use of an *equivalent* viscous damping model—derived from a nonlinear model and represented in terms of design variables in an explicit manner—is proposed.

The analysis of the hydraulic part of the injector resulted in the definition of an *equivalent* hydraulic scheme, on which basis both the equations of continuity in chambers and flow through nozzles were written.

fold

The peak value is 19-*fold* greater than the mean base line value.

Genes associated with DNA repair were mostly up-regulated in the naïve B cell to CB transition, including the mismatch repair genes *PMS*2 (5-*fold*), *MLH*1 (6-*fold*), and *MSH*6 (3-*fold*).

The scope of this control activity has increased many *fold* during recent years as the result of the rapid growth in types and size of military power plants.

in line with

Here we demonstrate that a severe functional disruption of the heart was produced by environmental factors. The results are *in line with* our previous finding regarding heart health and nutrients.

Test results show the LP turbine performance to be *in line with* expectation.

match

It is generally accepted that if sequences from the genome can be *matched* to sequences of proteins of known structure, we can usually determine whether they are related.

This experiment was done by training bees to *match* a sample pattern with one of two comparison patterns that they encountered subsequently in a decision chamber.

Specification of a turbocharger for a given engine involves *matching* the turbocharger performance characteristics with those of the piston engine.

Once the IBC engines for the candidate gas turbines were designed, an analysis has been developed to check the possibility to *match* these engines with other gas turbines, similar to those for which the IBC engines have been designed.

resemble

The best electron coupling routes in proteins *resemble* those in alkane bridges.

Te most notable facts about those b-waves are that they closely *resemble* each other but are not identical.

In theory, single-stage vehicle operations would *resemble* aircraft operations where high initial development costs are offset by relatively low recurring costs.

At higher frequencies, the directional characteristics r*esemble* that predicted by linear theory for a piston source.

respectively

The temperature increase at 1, 2, and 4 hours was 0.5, 2, and 4 degrees, *respectively*.

A has 16 nM affinity for receptor A, but is only 10-, 20-, and 30-fold selective over C, D, E receptors, *respectively*.

MCPI is about 90 and 700 times faster than MCNP5 for 2 and 1-mm^3 voxels, *respectively*.

The temperature near the exit of the combustor reached 1550 K, when the mass flow rate and fuel/air equivalence ratio were 0.06 and 0.8 g/sec, *respectively*.

similar

The curve for all glaciers is notably *similar* to the curve for glacier of the northern hemisphere.

Electron tunneling across a single THF molecule should have a decay factor *similar* to that of an alkane bridge.

An analysis has been developed to check the possibility to match these engines with other gas turbines, *similar* to those for which the IBC engines have been designed.

The component sizes are *similar* to those under consideration for the solar-powered Space Station, but the models can easily be generalized for other applications with axial or mixed-flow turbomachinery.

versus

We measured the activity of the conjugate molecules *versus* their individual components.

The selectivity of compound 3 decreased 4-fold *versus* compound 1.

Waterfall acceleration spectra *versus* rotor speed show the effects of increasing lubricant inlet pressure and temperature on turbocharger rotordynamic response.

Comparisons of active *versus* passive noise control techniques can be found in various papers.

4.11 用于讨论和描述实验意义的词汇

用于讨论和描述实验意义的词汇，常用于 Discussion 部分。

accompany

This crystal showed an exotic M-1 phase transition *accompanied* by a large structure change.

Metabolic changes *accompanying* brain activation do not appear to follow exactly the time-honored notion of a close coupling between blood flow and the oxidative metabolism of glucose.

account

The high frequency of DNA mutations may *account* for the accumulation of somatic mutations in mtDNA.

In earlier modeling studies, they did not take into *account* seasonal variation of the forcing which is a major factor in the simulated climate changes.

Based on the common definition for the Reynolds number an adjusted artificial Reynolds number has been derived taking into *account* the compressor mass flow and blading geometry

The engine's idling duration is taken into *account* but not plotted.

anticipate

It was *anticipated* that tumor cells would respond to such an intense therapy.

It is *anticipated* that such systems will play a key role in future deployments.

Some of these generator designs are *anticipated* to have electromechanical conversion efficiencies between 40% and 50%.

If a component cannot be made directly resistant to *anticipated* wear, sacrificial components can be placed at key locations.

argue

This would *argue* that B23P and B27W might exhibit different activities.

Many conservation biologists *argue* that the global application of wildlife-friendly farming methods would reduce the impact of agriculture on biodiversity.

attribute

It is reasonable to *attribute* observed dynamic vibration in the time domain to the coherent photo-like vibration.

This luminescence quenching is *attributed* to electron transfer from *D to A.

Central bursting has been *attributed* to material structural damage.

aspect

We report cerebral blood flow activity in profoundly deaf signers processing specific *aspects* of sign language in key brain sites widely assumed to be unimodal speech or sound processing areas.

Highly excited chemical species in many *aspects* of science and technology were carried out in extreme environments, including combustion, atmospheric, and interstellar phenomena.

After the discussion of all the technical *aspects* of the clutch strategy, the consumer acceptance has to be discussed as well.

The design of these elements constitutes an important *aspect* of mold design.

become

With the advent of atomic force microscope it *became* possible to study individual sliding junctions at the molecular level.

It was proven to be reproducible, reliable and accurate, and *becomes* progressively recognized worldwide as part of the Quality Control (QA) procedures for new beams.

Whereas in the recent years power density and reliability where the main development goals, the set of requirements has *become* considerably wider.

When this value *becomes* very small for angular variation, the cycle number will increase unnecessarily.

conceive

It is *conceivable* that the same mechanism accounts for the delayed relaxation of the surface pH changes into the bulk observed previously with bacteriorhodopsin membranes.

It is illogical, and directly opposed to the workings of evolutionary force, to *conceive* of a wide-spread group of animals suddenly appearing and springing into prominence.

Calcimimetic compounds could *conceivably* provide a specific medical therapy for primary hyperparathyroidism.

emphasize

He *emphasized* that the currently rich countries have agricultural and institutional needs that differ importantly from those of the currently poor countries.

The present study *emphasizes* the importance of cell surface expression and secretion of heparanase in tumor angiogenesis and metastasis.

The elasticity tests *emphasize* the full-load torque:

t60/100 km = h in the second-highest gear

t80/120 km = h in the highest gear

We *emphasize* that the design variable associated with X is still deterministic even though X is random.

exploit

The effects of human population growth on weather were *exploited*.

The uranium–238 and 235 isotopes are chemically identical, and can only be separated using techniques that *exploit* the differences in their atomic masses.

In reality the potential is hardly *exploited*, as drivers tend to drive in higher gears and at lower load levels due to the delayed response of turbocharged engines.

focus

We *focus* here on two particular data sets that allow us to address the evolution of deformation with time.

Previous studies have *focused* on preventing unified coherent motion of a swarm.

This thesis *focuses* on demonstrating a way to improve the responsiveness of turbocharged SI engines.

Most efforts have been *focused* on developing process models and computer-simulation tools for predicting LCM process behaviors.

given

Given the growing scale and impacts of agriculture, how should we best to resolve the need for increased food production with the desire to minimize the impact on wild life?

This question is particularly important *given* that the fundamental unit of infectivity remains unknown.

Full details of these tests and of the associated data processing are *given* in Ref. [18].

A formula was given for calculating the required sum of cutter numbers dependent on a *given* length to the tangency zone when cutting teeth with equal height teeth without correction of the generating motion.

hypothesize

Because vitamin C is present in high concentrations in the pituitary, we *hypothesized* that vitamin C might play a role in the secretion of anterior pituitary hormones.

Genetic and phenotypic instability are hallmarks of cancer cells, but their cause is not clear. The leading *hypothesis* suggests that a poorly defined gene mutation generates genetic instability and that some of many subsequent mutations then cause cancer.

imply

This data *imply* that both the pattern and strength of association may vary as the intrinsic cell state changes.

The spontaneous coherence *implies* that individual agents do not need to be manually controlled. Indeed, that is our main goal of such research.

Our results may also have *implications* concerning the fundamental limits of computation devices.

Commercial equipment is identified for completeness and does not *imply* endorsement by the authors.

in the presence of/in the absence of

The receptor-ligand binding kinetic parameters were observed to alter dramatically *in the presence* of Zn^{2+} but not other divalent cations.

The transient-state kinetics of binding of myosin to actin *in the presence* and *absence of* Ca^{2+} were investigated.

In the current investigation, *in the absence of* a universally applicable heat transfer model for diesel engines, a simple method proposed by Walker is used as a first approximation.

Therefore, *in the presence of* large rotational masses such as pulleys, harmonic balancers, flywheels or flex plates, the impact of coupled rigid and flexible body dynamics can be significant and hence needs to be considered.

indicate

Spectral analysis *indicated* the presence of drug compound.

Data from this study *indicated* that 5-HT receptor may play a role in normal memory. However, interpretation of these results has been somewhat controversial.

The irregular path plotted for the opening phase *indicates* that the present consideration of the closing phase is independent of the path followed in the opening phase.

A positive FET value *indicates* an ejection force is necessary.

insight

The Human Genome Project provides powerful new *insights* into human diseases and raises many challenging questions.

To gain *insight* into the structural basis of DNA bending by adenine-thymine tracts, we have determined the crystal structure of the high-affinity DNA target of the cancer-associated human papillomavirus E2 protein.

This simplified approach to the engine cycle offers some valuable *insights*, but is no replacement for more complex models.

Having a reliable and fast numerical model will help to achieve additional *insight* into the process itself and to avoid numerous time consuming and expensive experiments.

interpret

Most of the experimental results can be *interpreted* in terms of the electronic properties of the materials.

In the usual psychological *interpretation*, stimulations are designed to activate specific mental processes identified by cognitive psychology, which are then localized by the signals in functional imaging experiments.

The correct *interpretation* is that the same amount of heat is transferred to the fluid as it is discharged from the fluid.

Through this formalization, a first-order quantile *interpretation* of inverse FORM is also given.

notion

In 1950, John Nash contributed a remarkable one-page PNAS article that defined and characterized a *notion* of equilibrium for *n*-person games. This *notion*, now called the "Nash equilibrium," has been widely applied and adapted in economics and other behavioral sciences.

Several lines of evidence support the *notion* that the brain exhibits a significant degree of experience-dependent functional plasticity even in adulthood.

It is based on the *notion* that if one active element in a vibration transmission path can provide reasonable vibration attenuation, then two active elements in series may provide even better advantages and control possibilities.

The invariant definition of nonradiating steady-structure flow has been introduced as a fundamental extension of the ordinary *notion* of steady flow.

predict

To test the above *predictions*, we used the experimental set-up shown in Fig.3.

The human population is *predicted* to rise to between 8 and 10 billion by 2050.

A new, simple, and comprehensive agility index is thus developed, rating the wish for a fast torque development as well as the expectation of a *predictable* and comfortable response to the variation in accelerator pedal position.

For optimal balancing and diagnostic purposes it is important to be able to correctly *predict* the system vibration behavior over the operating speed range.

propose

Based on the kinetic study, a reaction model was *proposed*.

We *propose* that the effects we observed are caused by a new scattering channel.

The speed can be calculated from the torque equivalence when the vehicle is cruising at the *proposed* speed of 60 km/h.

Various techniques of metal forming process design have been *proposed* in the literature.

prove

Identification of the structure of Neurocin has *proven* very challenging.

In contrast, the R-compound *proved* to be potent agonist.

Such experiments served first to *prove* in principle the applicability of the proposed identification techniques.

provide

Extensive investigation of long-range electron transfer in proteins *provide* compelling support for the notion that electronic coupling pathways involve both covalent links as well as van der Waals and hydrogen-bonded contacts between strands of the polypeptide.

Ecosystem services are critical to human survival; in selected cases, maintaining these services *provides* a powerful argument for conserving biodiversity.

This approach *provides* poor agreement, as the speed lines intersect at an acute angle with the iso-pressure lines.

Without an accurate estimation of the fiber permeability profile, simulation models would not *provide* correct information, which is useful for process design.

reflect

This founding might *reflect* a race difference, rather than an age difference.

The animal population in this area likely *reflects* the influence of its geological and ecological factors.

The model *reflects* the real physical process, and the iteration calculation can be eliminated, which is vital to the hardware-in-the-loop simulation.

An activity-based approach is employed in order to better *reflect* the impacts of sacrificial components on remanufacturing costs.

shed light on

The structure of FAS *shed light on* its activity and fatty acid chain selectivity.

We present experiments that *shed light on* the basic mechanisms of light-matter interactions.

suggest

Both experiments *suggested* that chronic lack of vitamins accelerated tumor growth.

Maintenance of the biological activity in the presence of its antagonist *suggests* the presence of another receptor for PYY.

These results *suggest* that the current method for configuration state identification is practically feasible and useful though further investigation is needed to establish the viability of measuring the slope sufficiently accurately.

This *suggests* that in conventional shearing, very small clearances may provide an advantage of less tool sharpening.

support

Phylogenetic analyses of these and other RNA polymerase sequences strongly *support* the notion that Microsporidia are not early-diverging eukaryotes but instead are specifically related to Fungi.

The results *support* a mechanism in which cooperative interactions occur even in the relaxation process.

Turbogenerators used for power generation consist of several rotor stages that are rigidly coupled together to form a single rotor *supported* on many hydrodynamic bearings, resulting in a statically indeterminate rotor-bearing-foundation system (RBFS).

via

A completely different approach was to control the system *via* normal vibrations of small amplitude.

The scattering might proceed *via* the electron-phonon interaction.

We can reduce the exhaust gas temperature, and thus reduce the NO_x emission *via* exhaust gas recirculation (EGR).

The displacement of the rack bar is read into the Control PC *via* the position displacement sensor and converted to the appropriate fuel flow rate.

用于描写可能、猜测、好像等的词汇。

appear

Both MEP synthase and keto-acid reductase *appear* to show similar stability.

The glass formed at high pressure and temperature *appears* to have a superior thermal stability when compared with alloys formed from the conventional melting-cooling process.

Preliminary results indicate that errors in configuration state identification using the current approach are of the order of 20 μm, whereas those obtained using the Reynolds equation *appear* to be significantly higher.

Hence, it *appears* that it is possible to achieve our prestated objective of detecting DOC variations of over 10% with estimation errors of 15% or less on the average.

assume

We *assume* that the upper mantle layer has a viscosity three times greater than the asthenosphere below.

Our observed energy is a factor of 300 below that *assumed* by Smith.

It is *assumed* that one has an adequate model of the rotor, which is axially symmetric and can be adequately represented by system matrices.

If these estimated DOC variations are within ±10% of the nominal, we will *assume* they are "normal," as discussed earlier, and no fault of this nature is detected.

perhaps

Perhaps long ago, two nearly spherical Kuiper belt objects of comparable size coalesced to form a compound object.

Induction of many analyzed *rec* genes requires Rep1, *perhaps* in a complex with Cdc10, a transcriptional activator that regulates the mitotic cell cycle.

Material limits in the silicon may allow operation with inlet air temperatures as high as 950 K and *perhaps* 1000 K.

The low value of friction coefficient is *perhaps* due to lubrication at the contact surfaces.

presume/presumably

The *presumed* drop of oxygen levels would have dramatic effects on the biopopulation at that time.

Finally, the preference of formation of the phenol 3b *presumably* was the result of water contamination during the acid-catalyzed rearrangement.

Adding a fatty acid substitution would *presumably* increase dramatically the chance to cross the blood brain barrier.

The working fluid, *presumably* steam, flows through the circuit in Fig.1(a) denoted by the state points 1, 4, 5 and 7.

seem

The technique described in the paper *seems* to be useful for achieving this purpose.

The authors offer a thoughtful and elaborate model for future training in graduate clinical chemistry. However, they interpret history and present trends in a manner that *seems* to ignore the realistic demands of economic forces.

This parameter does not *seem* to be greatly affecting the injector behavior. However, a faster injector response can be seen when the control volume is reduced.

Thus, it would *seem* that furnace redesign is the key to providing the level of diameter control needed in emerging specialty optical fibers.

speculative

Even if the cause of AD is still *speculative*, aggregates are thought to be mainly responsible for the devastating clinical effects of the disease.

Although *speculative*, our technique might be useful for telecommunication.

What follows is in some sense *speculative*, but we feel it offers the most likely way forward if specialty fibers are to be produced with a diameter variation within the touted ±0.1 μm requirement.

suppose

Suppose that F is the same as B, then $X=F/B=1$.

Thus, *suppose* once again that unskilled labor is used only in the import-competing sector of the economy and skilled labor is only used in the export sector. How could this economy absorb an increase in the supply of unskilled workers if world commodity prices remain unchanged?

This in itself has significant implications for a process that is *supposed* to be controlling the measured diameter of telecommunications fiber to better than ±1 μm.

4.12 常用的修饰词

apparent

This *apparent* discrepancy is due to the fact that many data is not updated.

A proper treatment of the ray equations in the uncoupled azimuth approximation can also be shown to contain an azimuthal deflection angle which would produce an *apparent* horizontal refraction in a 2-D calculation.

approximately

The two kinds of electrodes gave potential measurements differing by *approximately* 15 mV.

The ratios of the molecular weights of these species are *approximately* 4:2:1.

The maximum displacement obtainable using an IR laser diode is *approximately* 30 microns.

clearly

DNA sequence data *clearly* support the inference that the Betsiboka River serves as boundary between two different *Lepilemur* species.

Measurements show *clearly* that walking on a wooden floor produces low frequency sounds which are difficult to measure and rate in any case.

compelling

Land-use and land-cover changes has received less attention, compared to liquid flows, although there is *compelling* evidence that such alterations can influence the functioning of the Earth System.

completely

The virus was *completely* eliminated after one course of drug therapy.

The measurement results show a good correspondence with the calculations and *completely* confirm the high expectations set in the measures.

considerable

These 6-deoxytetracyclines are *considerably* more resistant to degradation than their 6-hydroxy counterparts.

The maximum deviation of the side force between quasi static and dynamic model can be *considerable*.

critical

To better understand the binding interaction between antigen and antibody we need to distinguish protein residues *critical* to the binding energy and mechanism from residues merely localized in the interface.

For this purpose the prototype engine structures are excited internally at *critical* internal excitation locations and the noise emission at a distance of 1metre in free field is measured.

crucial

Deep understanding of the nucleation process in terms of physics is *crucial* for the short-term earthquake prediction.

It was found that the flow round the leading edge of the windward side of the rectangular cylinder plays a *crucial* role in the generation of the self-excited sound, but the formation of separation bubble there is not a necessary and sufficient condition.

dramatic

It appears to be true that substitution at position 5 has a *dramatic* effect on the stability of this series of derivatives.

essential

The interaction of an organism with its environment is *essential* to its survival.

315 genes are identified as *essential* for early zebrafish development.

exactly

DNA sequencing was performed *exactly* as described previously.

The actual environmental systems cannot be *exactly* modeled only by such a simplified regression model.

exclusively

Northern and Western blot analyses, using the *c*DNA and antibody, revealed its expression *exclusively* confined to the kidney.

extensively

Higher order Lagrangian theories have been studied *extensively* in recent years.

The crankshaft and engine block dynamic behavior has been *extensively* studied using mainly the finite element method due to its flexibility in modeling complex geometries.

extremely

The rapid and sensitive determination of pathogenic bacteria is *extremely* important in medical diagnosis and public health.

Acoustic field generated in real ship conditions is *extremely* difficult to describe by means of theoretical methods.

highly

These genes are *highly* expressed in ES and EC cells but not other differentiated tissue types.

However, such complete treatments of the propagation problem can still be *highly* inefficient.

intense

Genomic research has created *intense* interest in the use of microarray technology.

Blade-vortex-interaction (BVI) can be an *intense* source of rotorcraft noise.

interestingly

Interestingly, hormone insensitivity in the root response to ethylene was apparent in each of the mutants.

Interestingly, the fluctuations of beam intensity did not completely correlate with the mean intensity, i.e. some low noise beams had higher fluctuations than higher noise beams.

likely

These species are *likely* to become threatened in the future.

The process of machining a work piece usually involves a variety of cutting data which in turn are *likely* to cause substantial variations in the spectral properties of the tool vibrations.

nearly

Some stars exist which may be *nearly* as old as the atoms they contain, yet other stars must have been recently born.

The Zwicker method can be regarded as *nearly* linear.

nicely

The atomic model was *nicely* fitted into the density map at 7-A resolution obtained by electron cryomicroscopy.

Their biological activity correlated *nicely* with their in vitro receptor binding activity.

permanently

Unfortunately, molecules longer than 1-2 Mbp can become *permanently* immobilized or trapped after traveling various distances through the gel.

Gearbox shaft vibrations were simultaneously recorded from the *permanently* installed proximity probes as the compressor speed was varied over the speed range.

potentially

The U.S. Environmental Protection Agency (EPA) recently report that 189 *potentially* hazardous air pollutants are emitted from coal-burning electric utility generators.

Note that this need to measure slopes is *potentially* a significant shortcoming of this alternative identification technique.

precisely

Animal behaviors are controlled by a complex, *precisely* timed interaction of several neuropeptides.

It did not identify *precisely* the local dynamics ultimately responsible for the generation of radiated sound.

predominantly

HuR appears *predominantly* nucleoplasmic but has been shown to shuttle between the nucleus and cytoplasm via a novel shuttling sequence.

This is *predominantly* due to the presence of exhaust and engine carcass noise in the case of diesel locomotives, and cooling fan noise for both diesel and electric locomotives.

probably

The actual population in this area in the 1100 s is of course unknown, but was *probably* lower that that of the present-day population.

The slightly poorer identification results when using the Reynolds equation are *probably* due to the increased data processing requirements.

profound

Regulation of these proteins may have *profound* effects on carcinogenic activity.

It is demonstrated that the active control has a *profound* enhancement on the transmission performance of the plate at low frequencies.

radically

We expected that such a compound would exhibit a *radically* different pharmacological profile from the known molecules.

readily

These polymers are made of the *readily* available monomers.

Upon normalization, the vibration equations can be expressed in terms of natural frequencies and modes, from which a state-space model can be *readily* obtained as a control model.

reasonable

The latter technique would require more precision than we can presently achieve at *reasonable* costs.

It is concluded that the method provides a *reasonable* balance between ease of calculation and accuracy for some applications, but for some specific configurations anomalous results are observed.

roughly

The decline of drug concentration is *roughly* exponential and occurs with a half-life of around 11 hours.

A general theory for finding a precise regression relationship between these two kinds of fluctuations is proposed through the *roughly* observed data.

spontaneously

Upon addition of preformed lipid vesicles, the protein *spontaneously* refolded and inserted into the vesicle membranes.

strongly

Analyses of these and other RPB1 sequences *strongly* support the notion that Microsporidia are not early-diverging eukaryotes but instead are specifically related to Fungi.

It was noticed that the above-mentioned assumptions make the model *strongly* non-linear with clear interaction between torsional and radial vibration.

substantial

There is *substantial* experimental evidence that Xepril on administration become associated with BSA.

While the predicted change in ignition delay when boost pressure is reduced to 1.65 bars is relatively small, the peak RHR value during the premixed phase of burning is *substantially* higher.

surprisingly

Surprisingly, even in the absence of Ku, p460 was capable of binding to DNA and being activated.

temporally

This pattern of results supports the view that predictable reward presentations *temporally* organize search states and related responses between food presentations.

thoroughly

Blood cells were separated, washed *thoroughly*, and tested immediately after collection.

There is no evidence that this topic has been *thoroughly* studied.

virtually

Substitution by a chloro group has *virtually* no effect on the receptor binding selectivity.

附：稿件样本

A sample manuscript

Characterization of the Antioxidant Activities of Danshensu and Salvianolic Acid B in Chemical Systems and Human Endothelial Cell Culture

Guan-Long Zhou[a], Hen-Mi Zhang[a], Tin-Xian Ye[a], Shi-Jun Xiang[a], Yin-Lin Duan[a*], Zhi-Win Guo[b]

[a] Department of Pharmaceutical Chemistry, School of Chemical Engineering and Technology, Yanjin University, Yanjin 342002, P.R.China

[b] Research Institute of Sciwriting Co. Ltd., Jinan 250100, P. R. China

[*] Corresponding author. Tel: +86-12-7403798; Fax: +86-12-2742688.

E-mail: bosi@sciwriting.cn

Abstract

Salvia miltiorrhiza is one of the most widely used traditional Chinese medicine for the treatment of various cardiovascular diseases. Danshensu [3-(3, 4-dihydroxyphenyl) lactic acid] and salvianolic acid B, two phenolic acids isolated from salvia miltiorrhiza, protected animals and cells from oxidative injury. To better understand their biological functions, the in vitro antioxidant activities of danshensu and salvianolic acid B were evaluated along with vitamin C in different chemical systems. Both danshensu and salvianolic acid B exhibited higher reducing power in the Ferric to Ferrous reductive reaction, and scavenging activities against free hydroxyl radicals, superoxide anion radicals, 1,1-diphenyl-2-picryl-hydrazyl (DPPH) radicals and 2-azino-bis (3-ethylbenzthiazoline-6-sulfonic acid) (ABTS) radicals than vitamin C. In contrast, danshensu and salvianolic acid B showed weaker iron chelating and hydrogen peroxide scavenging activities than vitamin C. Pretreatment with salvianolic acid B and danshensu inhibited hydrogen peroxide induced endothelial cell damage. The protective efficiencies were correlated with their antioxidant activities. These results indicated that danshensu and salvianolic acid B are efficient antioxidants and salvianolic acid B is superior to danshensu. Their antioxidant properties might contribute to the pharmacological functions of Salvia miltiorrhiza and suggest that they could have potential applications in food industry.

Keywords: Danshensu; salvianolic acid B; antioxidant; radical scavenging; DPPH; ABTS radical; human vein vascular endothelial cell

Introduction

There is increasing interest in natural antioxidant products for use as medicines and food additives[1]. Vitamin C, vitamin E and carotenoids are some of these widely used natural antioxidants. Antioxidants played an important role in lowering oxidative stresses caused by reactive oxygen species (ROS). ROS including superpxide anion radical, hydroxyl radical and hydrogen peroxide are generated under physiological and pathological stresses in human body[2]. Some severe chronic diseases, such as arthritis, cancer, diabetes, cardiovascular diseases, inflammations and neurological disorders are related to the imbalance of ROS formation and their elimination[3]. Antioxidants play an important role in protecting against these complex diseases through scavenging free radicals and reducing hydrogen peroxide[4]. Therefore, the identification and study of novel compounds characterized with antioxidant activity from natural sources is an important strategy to improve human health condition and life quality.

Salvia miltiorrhiza (Danshen in Chinese) is one of the most widely used traditional herb medicines for the treatment of a variety of diseases, such as cardiovascular diseases, hepatitis, hepatocirrhosis, chronic renal failure and dysmenorrhea[5]. The formulae derived from this herb like Compound Danshen Dripping Pill, Danshen Pian and Danshen Injection have been developed and used in the clinic in China, Korea and Russia[6]. The pharmacologically active compounds of Salvia miltiorrhiza comprise two fractions: lipophilic diterpenoid tanshinones and water-soluble phenolic acids. In recent years research interests have been focused on phenolic acids. Total twenty-five phenolic acid compounds have been isolated and identified from this species[7], of which danshensu[3-(3, 4-dihydroxyphenyl) lactic acid]and salvianolic acid B (lithospermate acid B) (Figure 1) have the highest contents, accounting for over 1% and (3-5) % of total dried weight, respectively[8].

Etc.

Materials and Methods

Chemicals

2, 2'-Azino-di-3-ethylbenzthiazoline sulphonate (ABTS), 2, 2'-azobis (2-amidino-propane) dihydrochloride (AAPH), 3-(2-pyridyl)-5, 6-bis(4-phenyl- sulfonic acid)-1, 2, 4-triazine (Ferrozine), 1, 1-diphenyl-2-picrylhydrazyl (DPPH), nicotinamide adenine dinucleotide (NADH), Nitroblue tetrazolium (NBT), phenazine methosulfate(PMS), 1, 10-phenanthroline, potassium hexacyanoferrate [$K_3Fe(CN)_6$], 3-(4, 5-dimethylthinazol-2-yl)-2, 5-diphenyl tetrazolium bromide (MTT), 30% hydrogen peroxide (H_2O_2), and vitamin C (L-ascorbic acid) were purchased from the Sigma Chemical Co. (St. Louis, USA). Salvianolic acid B and danshensu, isolated from dried roots of *Salvia miltiorrhoza* Bunge, were obtained from the China National Institute for the Control of Pharmaceutical and Biological Products (Beijing, China).

Reducing Power Assay

The Ferric reducing power assay was performed as previously described by Oyaizu (*29*) with slight modifications. The reaction solution was freshly prepared by mixing A, B, C in the ratio of 1∶1∶1 (*V/V/V*), where A is 0.2 M phosphate buffer (pH 6.6), B is 1% (w/v) potassium hexacyanoferrate [$K_3Fe(CN)_6$], C is various concentrations of test compounds. The mixture was incubated at 50 ℃ in a water bath for 20 min before adding 10% (*W/V*) trichloroacetic acid (TCA) solution to stop the reaction. The supernatant collected by centrifugation at 3000 r/min for 10 min was mixed with equal volumes of distilled water and ferric chloride ($FeCl_3$) solution to a final $FeCl_3$ concentration of 0.1 mM. After 10 min incubation at room temperature, absorbance at 700 nm was measured. A higher absorbance indicated a greater reducing power.

Cell Culture and Cell viability Assay using MTT

Human umbilical vein endothelial cells were isolated from freshly obtained human umbilical cords by collagenase type Ⅱ treatment as previously described [35]. Endothelial cells were counted and seeded at a density of 5×10^4 cells/well onto gelatin-coated 96-well plates containing endothelial growth medium supplemented with 15% fetal bovine serum, 100 U/mL penicillin and 100 μg/mL streptomycin, and cultured at 37 ℃ in a humidified environment containing 5% CO_2. After treatment with various concentrations of Salvianolic acid or Danshen for 4 h followed with 0.4 mM H_2O_2 for 18 h, each well was washed twice with PBS to remove the medium. One hundred microliter of MTT (0.5 mg/mL) was added to each well and incubated at 37 ℃ for an additional 4 h. Finally, 150 μL dimethyl sulphoxide (DMSO) was added to each well and the absorbance at 570 nm was read. The absorbance was used as a measurement of cell viability. It was normalized by the absorbance of cells incubated in control medium, which were considered 100% viable.

Etc.

Results

Ferric Reducing Power

The reducing powers of salvianolic acid B and danshensu were examined by using the Fe^{3+} to Fe^{2+} reductive reaction. As shown in Figure 2, the reducing powers of salvianolic acid and danshensu were concentration-dependent and increased with increasing concentrations. Salvianolic acid B showed significantly higher reducing power than vitamin C at all tested concentrations. In addition, the reducing power of danshensu was slightly higher than that of vitamin C at a given concentration, but their differences are not significant. These results indicated that both salvianolic acid B and danshensu are strong reductants and salvianolic acid B is the stronger one.

Ferrous Ion Chelating Activity

Chelating agent can interfere with the formation of ferrous and ferrozine complex and thus decrease the red color. The chelating activities of the salvianolic acid B and danshensu for ferrous ion were assessed. As shown in Figure 3, salvianolic acid B and danshensu slightly inhibited the formation of the red-colored complex. At maximum concentration of 1.0 mg/mL, salvianolic acid B and danshensu chelated merely 7.5% and 2.8% of ferrous ions, equivalent to approximately 1/2 and 1/5 of that of vitamin C, respectively. These results suggested that salvianolic acid B and danshensu are very weak Fe^{2+}-chelators.

Hydrogen Peroxide Scavenging Activity

The scavenging reaction included 100 μM H_2O_2 and various concentrations of test compounds. After a period of incubation, the remaining H_2O_2 was measured by titration. The hydrogen peroxide scavenging activities of Salvianolic acid B and Danshensu were shown in Figure 4. Salvianolic acid B and Danshensu exhibited some extent of hydrogen peroxide scavenging capacity. At the same concentration, Salvianolic acid B was a stronger scavenger than Danshensu. At a concentration of 100 μg/mL, the H_2O_2 scavenging activity of vitamin C was over 90%, but the scavenging activities of Salvianolic acid B and Danshensu were 30% and 20%, respectively. Both Salvianolic acid B and Danshensu showed a weaker scavenging activity against hydrogen peroxide as compared to vitamin C.

Etc.

Discussion

Phenolic acids are thought to play a positive role in the prevention of human diseases. Salvianolic acid B and danshensu are the most abundant active ingredients in salvia miltiorrhiza. Increasing reports on the protective effects of salvianolic acid B and danshensu against oxidative stresses implied that they may have antioxidant activities. An antioxidant exerts its antioxidant activity through various mechanisms, among them are reducing power, chelating of metal iron, decomposition of peroxide, and scavenging of free radicals. We speculated that the antioxidant profiles of salvianolic acid B and danshensu are similar, but may be quantitatively different, because they share similar but different chemical structures.

Etc.

Acknowledgments

This work was supported by grant 2006YCX027 from the natural Science Foundation of China and grant 2006BS06010 from the Young Scientists' Project of China.

References

[1] Willcox, J. K.; Ash, S. L.; Catignani, G. L.. Antioxidants and prevention of chronic disease. *Crit. Rev. Food Sci. Nutr.* 2004, *44*, 275-295.

[2] Nordberg, J.; Arner, E. S. J.. Reactive oxygen species, antioxidants, and the mammalian thioredoxin system. *Free radical Biol. Med.* 2001, *31*, 1287-1312.

[3] Halliwell, B.. Annual review: Antioxidants in human health and disease. *Nutrition*. 1996, *16*, 33-50.

[4] Shahidi, F.; Wanasundara, P.K.. Phenolic antioxidants. *Crit. Rev. Food Sci. Nutr.* 1992, *32*, 67-103.

[5] Li, L. N.. Biologically active components from traditional Chinese medicines. *Pure Appl. Chem.* 1998, *70*, 547-554.

[6] Zhao, G. R.; Xiang, Z. J.; Ye, T. X.; Yuan, Y. J.; Guo, Z. X.. Antioxidant activities of *Salvia miltiorrhiza and Panax notoginseng*. Food *Chem.* 2006 (in press)

[7] Jiang, R. W.; Lau, K. M.; Hon, P. M.; Mak, T. C. W.; Woo, K. S.; Fung, K. P.. Chemistry and biological activities of caffeic acid derivatives from *Salvia miltiorrhiza. Curr. Med. Chem.* 2005, *12*, 237-246.

Etc.

Figure legends

Figure 1. Chemical structures of caffeic acid (**1**), danshensu (**2**) and salvianolic acid B (**3**).

Figure 2. Ferric reducing antioxidant power of Salvianolic acid (○), Danshensu (●) and vitamin C (■). Each value is the mean ±SD of five replicates ($n=5$).

Figure 3. Ferrous ion chelating activities of Salvianolic acid B, Danshensu and vitamin C. Each value is mean ±SD of five replicates ($n=5$).

Etc.

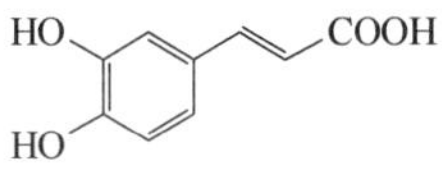

Figure 1.

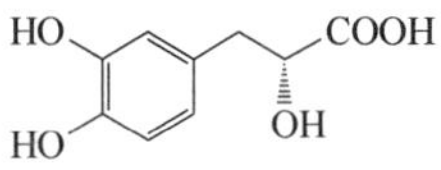

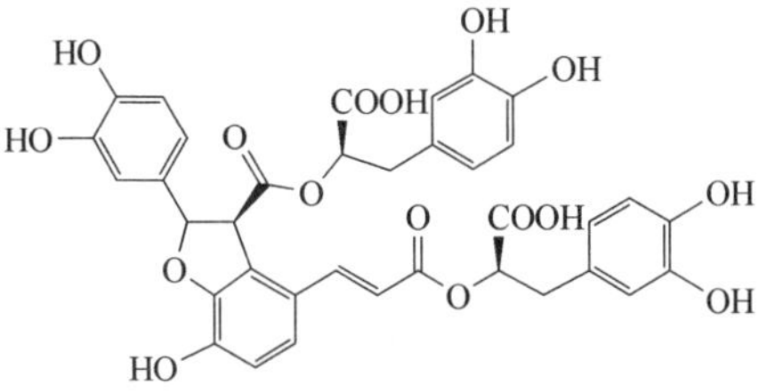

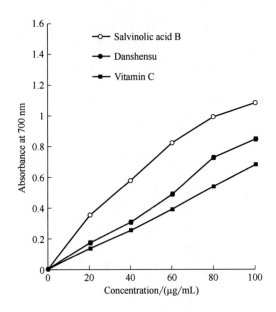

Figure 2.

附　录

1. Greek Alphabet

Name	Lower case	Upper case	English letter
alpha	α	A	A
beta	β	B	B
gamma	γ	Γ	G
delta	δ	Δ	D
epsilon	ε	E	E
zeta	ζ	Z	Z
eta	η	H	H
theta	θ	Θ	Q
iota	ι	I	I
kappa	κ	K	K
lambda	λ	Λ	L
mu	μ	M	M
nu	ν	N	N
xi	ξ	Ξ	X
omicron	o	O	O
pi	π	Π	P
rho	ρ	P	R
sigma	σ	Σ	S
tau	τ	T	T
upsilon	υ	Υ	U
phi	φ	Φ	F
chi	χ	X	C
psi	ψ	Ψ	Y
omega	ω	Ω	W

2. Prefixes for units of measure

Power of 10	Prefix	Symbol
10^9	giga	G
10^6	mega	M
10^3	kilo	k
10^2	hecto	h
10^1	deca	da
10^0	—	—
10^{-1}	deci	d
10^{-2}	centi	c
10^{-3}	milli	m
10^{-6}	micro	μ
10^{-9}	nano	n
10^{-12}	pico	p
10^{-15}	femto	f

3. Abbreviations

A.D.	anno Domini, after the beginning of the Century	cm	centimeter
		Corp	Corporation
a.m.	ante meridiem, before noon	cos	cosine
amp	ampere	Dr.	Dcotor
atm	atmosphere	Drs.	Doctors
B.A.	Bachelor of Arts	e.g.	exempli gratia in Latin, for example
Bbl	barrel		
B.C.	before Christ, before the beginning of the Century	Ed.D.	Doctor of Education
		et al.	et alii in Latin, and others
Btu	British thermal unit	etc.	et cetera, and so forth
B.S.	Bachelor of Science	F	Fahrenheit
©	copyright	fig.	figure
cal	calorie	ft	foot or feet
cd	candela	g	gram

gal.	gallon	μL	microliter
hp	horsepower	mL	milliliter
hr	hour	Mr.	Mister
Hz	hertz	Mrs.	Married woman
i.e.	id est, that is	Ms.	Woman of unknown marital status
in.	inch		
Inc.	incorporated	M.S.	Master of Science
Jr.	Junior	oz	ounce
kg	kilogram	Pa	Pascal
L	liter	Ph.D.	Doctor of Philosophy
lb	pound	p.m.	post meridiem, after noon
M.A.	Master of Arts	ppm	parts per million
M.B.A.	Master of Business Administration	®	trade mark
		s	second
M.D.	Doctor of Medicine	vol.	volume
μg	microgram	vs.	versus
mg	milligram	yd	yard
min	minute	yr	year

4. Journal name abbreviations

No abbreviation for single word title.

A

Abstract	abstr.
Academy	Acad.
Acta	Acta
Advances	Adv.
Agricultural	Agric.
American	Am.
Analytical	Anal.
Anatomical	Anat.
Animal	Anim.
Annals	Ann.
Annual	Annu.
Anthropological	Anthropol.
Antibiotic	Antibiot.
Antimicrobial	Antimicrob.
Applied	Appl.
Archives	Arch.
Association	Assoc.
Astronomical	Astron.
Atomic	At.
Australian	Aust.

B

Bacteriological	Bacteriol.
Bacteriology	Bacteriol.
Behaviour	Behav.
Biochemical	Biochem.
Biochemistry	Biochem.
Biochimica	Biochim.

Biological	Biol.	Edition	Ed.
Biology	Biol.	Electric	Electr.
Botanical	Bot.	Electrical	Electr.
Botany	Bot.	Endocrine	Endocr.
British	Br.	Engineering	Eng.
Bulletin	Bull.	Entomological	Entomol.
Bureau	Bur.	Environmental	Environ.
		Ethnology	Ethnol.
		European	Eur.
		Experimental	Exp.

C

F

Canadian	Can.		
Cancer	Cancer		
Cardiology	Cardiol.	Federal	Fed.
Cell	Cell	Federation	Fed.
Cellular	Cell.	Fish	Fish
Central	Cent.	Fisheries	Fish.
Chemical	Chem.	Food	Food
Chemistry	Chem.	Forest	For.
Chemotherapy	Chemother.		
Chimie	Chim.		

G

Chinese	Chin.		
Clinical	Clin.	Gazette	Gaz.
Communication	Commun.	General	Gen.
Computational	Compt.	Gene	Gene
Conference	Conf.	Genes	Genes
Contributions	Contrib.	Genetics	Genet.
Current	Curr.	Geographical	Geogr.
		Geological	Geol.

D

H

Diary	Diary		
Dental	Dent.	Helvetica	Helv.
Developmental	Dev.	History	Hist.
Diseases	Dis.	Human	Hum.
Drug	Drug		

E

I

		Immunity	Immun.
Ecology	Ecol.	Immunology	Immunol.
Economics	Econ.	Indian	Indian

Industrial	Ind.	**N**	
Institute	Inst.	National	Natl.
Internal	Intern.	Nature	Nat.
International	Int.	Natural	Nat.
J		Naturalist	Nat.
Japan	Japan	Neural	Neural
Japanese	Jpn.	Neurology	Neurol.
Journal	J.	Neuroscience	Neurosci.
K		Nuclear	Nucl.
Korean	Korean	Nucleic	Nucleic
L		Nutrition	Nutr.
Laboratory	Lab.	**O**	
Learning	Learn.	Occupational	Occup.
Letters	Lett.	Ocean	Ocean
Life	Life	Oceanography	Oceanogr.
M		Oceanology	Oceanol.
Magezine	Mag.	Official	Off.
Mammal	Mamm.	Ology	Ol.
Marine	Mar.	Opinion	Opin.
Material	Mater.	Organic	Org.
Mathematics	Math.	**P**	
Mechanical	Mech.	Paleontology	Paleontol.
Medical	Med.	Pathology	Pathol.
Medicine	Med.	Peptide	Pept.
Memoirs	Mem.	Pharmacology	Pharmacol.
Methods	Methods	Philosophical	Philos.
Microbiological	Microbiol.	Physical	Phys.
Microbiology	Microbiol.	Physiology	Physiol.
Molecular	Mol.	Physics	Physics
Monthly	Mon.	Plant	Plant
Morphology	Morphol.	Pollution	Pollut.
Mutation	Mutat.	Prevention	Prev.
		Proceedings	Proc.
		Progress	Prog.

Protein	Protein		
Publications	Pub.	Technology	Technol.
		Tetrahedron	Tetrahedron
Q		Theoretical	Theor.
Quarterly	Q.	Therapeutics	Ther.
		Tissue	Tissue
R		Transactions	Trans.
		Trends	Trends
Report	Rep.	Tropical	Trop.
Research	Res.		
Review	Rev.	**U**	
Royal	R.		
Russian	Russ.	United States	U.S.
		University	Univ.
S		Urological	Urol.
Scandinavian	Scand.	**V**	
Science	Sci.		
Scientific	Sci.	Veterinary	Vet.
Series	Ser.	Virology	Virol.
Service	Serv.	Virus	Virus
Society	Soc.	Vitamin	Vitam.
Solid	Solid		
Sound	Sound	**W**	
Special	Spec.		
Station	Stn.	Washington	Wash.
Structure	Struct.	Water	Water
Studies	Stud.	Work	Work
Surgery	Surg.	World	World
Survey	Surv.		
Symposia	Symp.	**Y**	
Symposium	Symp.		
Systematic	Syst.	Yeast	Yeast
		Z	
T			
Technical	Tech.	Zeitschrift	Z.
		Zoological	Zool.
		Zoology	Zool.

References

1. Robert Day. *How to Write and Publish a Scientific Paper*. Phoenix: Oryx Press, 1998.
2. George M. Whitesides. *Whitesides' Group: Writing a Paper*. Advanced Materials, 2004, 14, 1375-1377.
3. George D Gopen and Jusith A. Swan. *The Science of Scientific Writing*. American Scientist, 1990, 78, 550-558.
4. William Strunk, Jr. *The Elements of Style*. New York: New York Press, 1978
5. J. S. Dodd, *ACS Style Guide*: *A Manual for Authors and Editors*. 2nd ed.. Washington D C: American Chemical Society, 1997.
6. Proceedings of National Academy of Sciences.
7. Journal of American Chemical Society.
8. Journal of Chinese Pharmaceutical Sciences.